Download the TotalAR app to access enhanced features including updated NCEES *PE Civil Reference Handbook* page numbers for the solution references.

Scan this QR code with a smartphone or tablet to download the TotalAR app.

Unique Access Key:
YMPBXGF2

Certain features available through TotalAR targets are provided either on a complimentary basis, for a limited period of time, or for purchase, and in all cases are for the sole use of the original purchaser of this book. The unique code shall not be sold or otherwise transferred to any other person or entity. EduMind, Inc. and its School of PE division have the right to terminate access to any or all of these features for any reason and at any time with or without notice. EduMind, Inc. will make its best effort to resolve any technical issues as quickly as possible, but there will be no refunds, extended access periods, or other compensation for any technical issues, unavailability, or if the complimentary access is terminated for any reason. All content available through the TotalAR app is owned or licensed by EduMind, Inc. and subject to all applicable copyright, trademark, and other intellectual property laws and may not be copied, reproduced, or otherwise disseminated without written permission from EduMind, Inc. To request permission, email permissions@edumind.com.

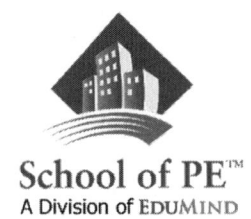

School of PE™
A Division of EduMind

www.schoolofpe.com

PE Civil: Structural Practice Exam & Solutions

EduMind, Inc. is not affiliated with the National Council of Examiners for Engineering and Surveying (NCEES) and all questions are not actual exam questions or questions provided by NCEES, but are similar in nature to the type of questions expected on the exam. All questions and provided solutions are not guaranteed to be error free and the sole intended use of the questions and solutions is for use as a study aid and not for any practical application.

For more details on the NCEES PE Civil exams, visit https://ncees.org/engineering/pe/civil-cbt/.

Copyright © 2022 EduMind, Inc.

All rights reserved.

EduMind, Inc. owns the copyright on all content, unless otherwise noted. No part of this book may be scanned, uploaded, reproduced, distributed, or transmitted in any form or by any means whatsoever, nor used for any purpose other than for the personal use of the original purchaser of this copy, without the express written consent of EduMind, Inc. To obtain permission, contact permissions@edumind.com.

April 2022

ISBN 978-1-970105-47-6

Printed in the USA

School of PE
An imprint of EduMind, Inc.
425 Metro Place N, Suite 450, Dublin, OH 43017
www.schoolofpe.com

Introduction

School of PE takes test preparation seriously, and we want users to pass their PE Civil exams. This *PE Civil: Structural Practice Exam & Solutions* volume contains features to help users gain an edge in preparing for the PE Civil: Structural exam.

A big part of preparing for the PE Civil exams is learning to navigate the NCEES *PE Civil Reference Handbook (PECRH)* and the approved codes and standards for each exam—the only resources test-takers can access during the exam. In most of this volume's solutions, you will see an NCEES *PECRH* section reference, a code reference, or a standard reference, directing the user to the relevant resources and sections that cover the question topic. (The questions that don't have a reference were deemed by our subject-matter experts to be basic knowledge for PE Civil test-takers.) You will notice that we did not include *PECRH* page numbers in the volume, as these change with each new version of the handbook. By looking for the necessary information in the *PECRH* and the approved codes and standards, you will learn how they are organized, saving you time during the exam. To find the page numbers of most of the references that appear in this book, use the augmented reality (AR) feature—*PECRH* Content Locator. The page numbers will be updated with the latest version of the handbook so you will always have the most up-to-date information handy.

To access this feature, download the TotalAR app (see the first page of this volume). This will allow you to gain access to the book's enhanced content. When you see a TotalAR (TAR) code, like the one for the *PECRH* Content Locator (p. iv), simply scan it with the TotalAR app to access the desired content. When you scan your first TAR code in the book, you will be prompted to enter a unique access key, which you will also find on p. i.

With this book, you also have access to an online version of this practice exam via the School of PE Exam Simulator, available on www.schoolofpe.com. If you purchased the book through our website, you should have received an email with a user ID and password. Once you log in, click the icon for this book to gain access to the online version of this practice exam. If you purchased the book from another site—for example, Amazon—please email your name and unique access key (from p. i) to info@schoolofpe.com to gain access to this exclusive online feature.

A Final Note

We have tried our best to make this book error free; it has been technically reviewed, edited, and tested. However, we are only human, and errors can happen. If you spot one, please notify us so we can improve the content. To report an error, scan the Feedback code (below) with the TotalAR app, or visit this feedback page:
https://publications.schoolofpe.com/books/pe-civil-structural-practice-exam/feedback.

We will verify the correct information and add it to the errata page, which will be made available immediately via the TAR code below and the following link:
https://publications.schoolofpe.com/docs/pecspes/v1/errata.pdf.

We welcome your feedback and want to hear how well this book prepared you for the PE Civil: Structural exam. If you have any suggestions or comments, please send them to us at publications@edumind.com.

Best of luck!
The School of PE team

      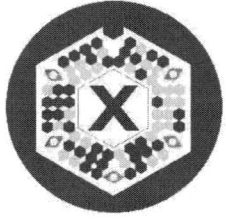

PECERH Content Locator   Feedback   Errata

PRACTICE EXAM

PRACTICE EXAM

1. The cross section shown below belongs to a 650-ft long concrete channel. The concrete thickness is 8 in. What is most nearly the volume of concrete needed to construct the channel given a 7.5% waste?

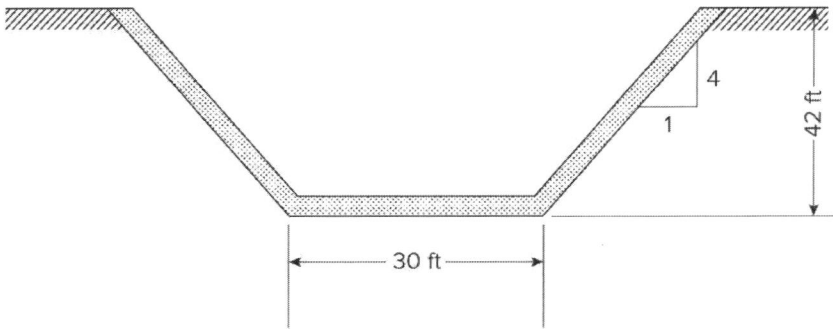

A. 1,265 yd³
B. 1,871 yd³
C. 2,012 yd³
D. 2,219 yd³

2. In the activity network below, what is the total float of task 5?

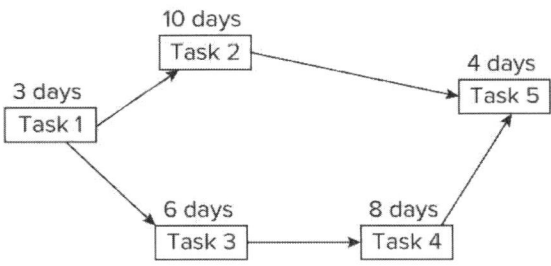

A. 0
B. 1
C. 2
D. 3

3. The owner of a proposed school would like to develop a 100,000-ft² building with enough parking to service staff members and parents. The budget for the project is $2,000,000. Based on previous projects, the owner has forecasted that each parking spot will cost $4,500, and every 90 ft² of building will cost $360 to construct. Assuming the entire building will be constructed, she will be able to afford _____ parking spots.

Fill in the blank.

4. An activity-on-arrow network is shown below. The variables are defined as follows:

TF = total float
LF = late finish
LS = late start
ES = early start
EF = early finish
D = duration

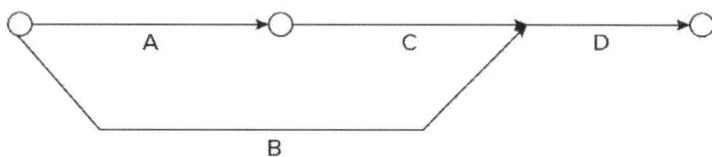

What is the expression for the total float of activity C (TF_C)?

    A. $TF_C = LF_D - EF_C$
    B. $TF_C = LF_C - ES_C - D_C$
    C. $TF_C = LS_C - EF_A$
    D. $TF_C = ES_D - ES_C - D_C$

5. A team has 15 days working 8-hr shifts to haul 182,212 yd³ of excavated loose material using a fleet of specified dump trucks that has a capacity of 20 yd³. The total cycle (load, haul, dump, and return) time is 40 min. The minimum number of trucks required is most nearly _____ trucks.

Fill in the blank.

6. A laborer is only able to apply a maximum force (P) of 50 lb applied perpendicular to the top of the wrench. The bolt will only be loosened at a torque of 750 in·lb along the vertical axis in the configuration as shown below. The minimum offset length (L) of the wrench that the laborer must select to loosen the bolt is most nearly:

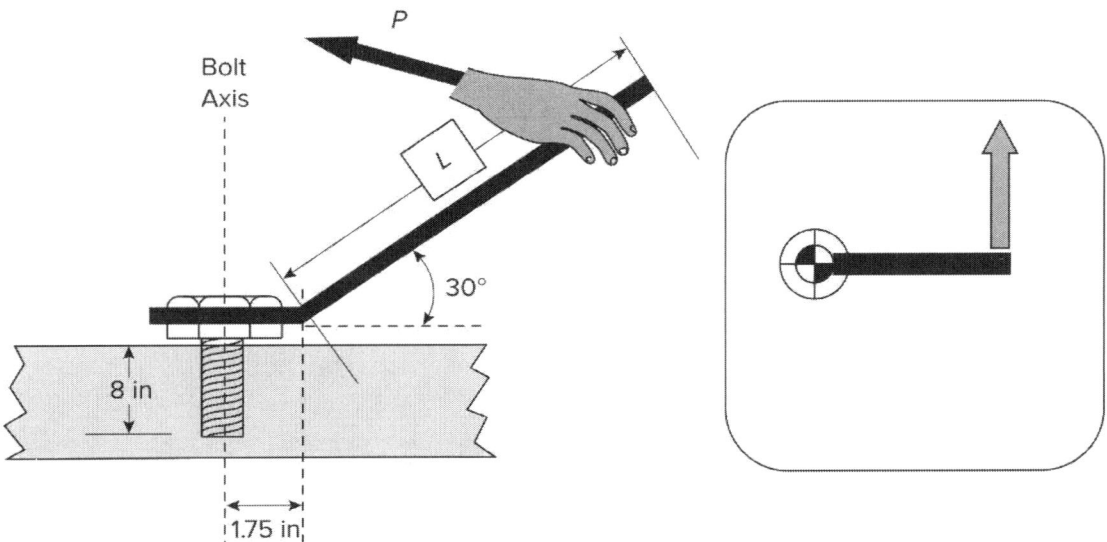

A. 13.25 in
B. 14.78 in
C. 15.3 in
D. 16.15 in

7. A formwork system for a cast-in-place wall is braced to support against possible wind loads as shown in the figure below. The connection at the base of the form panel can be considered a hinge. If each brace spaced 7 ft apart can support a maximum compressive axial force of 4,158 lb, the maximum wind load that can be supported is most nearly:

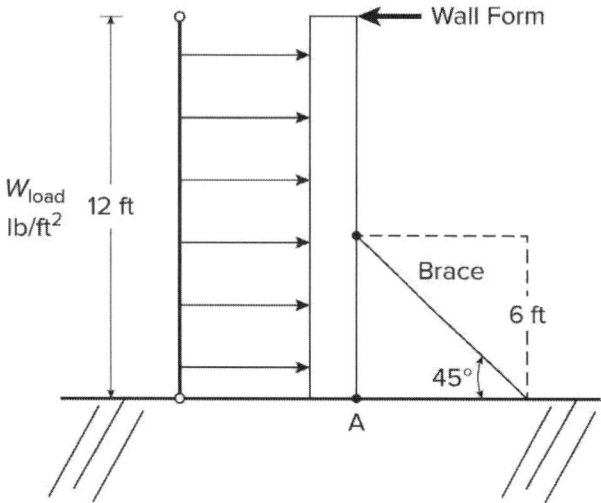

- A. 18 lb/ft²
- B. 30 lb/ft²
- C. 35 lb/ft²
- D. 40 lb/ft²

8. The following diagram shows a uniform soil deposit with information about each layer. Assume the effective stress at point A is 90 kPa. Determine the value of H_1.

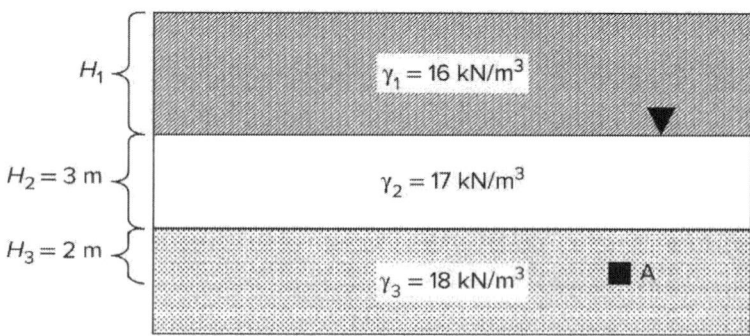

- A. 2.3 m
- B. 2.5 m
- C. 3.3 m
- D. 3.5 m

9. Point A sits in a sandy soil 8 m below the ground. The underground water level is 3 m below the ground. If the specific gravity (G_S) of soil solids is 2.7, the degree of saturation (S) is 0.5, and the water content of soil (ω) is 20% above the water level and 35% below the water level, determine the effective vertical soil pressure at point A.

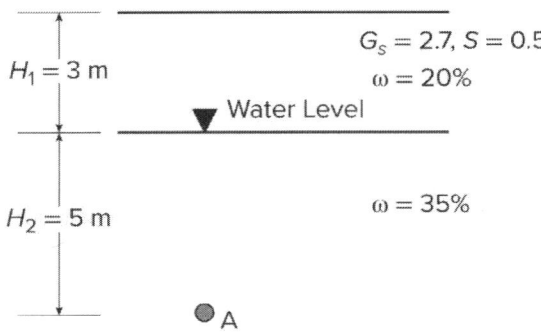

A. 23.5 kPa
B. 35.2 kPa
C. 53.2 kPa
D. 88.7 kPa

10. The following figure shows a 15-ft high frictionless wall. Assume failure along a plane oriented 60° from horizontal and a soil unit weight of 115 lb/ft³. Determine the active earth pressure per unit length of wall according to Rankine theory.

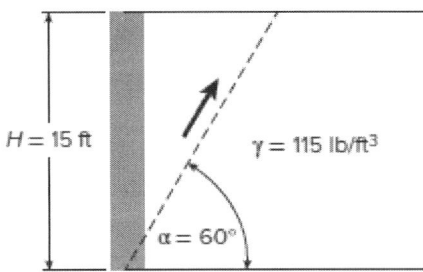

A. 850 lb
B. 4,270 lb
C. 8,540 lb
D. 13,830 lb

PRACTICE EXAM

11. A 20-ft soil sample is assumed to be in the range of virgin compression. Assume the initial effective pressure (p_0) is equal to 3,000 lb/ft², the effective stress increase (Δp) is equal to 2,000 lb/ft², and the sample experiences 12 in of settlement. If the initial porosity (n) is 0.5, what is the coefficient of consolidation (C_c)?

- A. 0.45
- B. 0.68
- C. 0.73
- D. 0.84

12. A 5-m soil sample is given. Assume the coefficient of reconsolidation (C_R) is given as 0.1 and the coefficient of consolidation (C_c) is 0.6. The initial effective pressure (p_0) is equal to 300 kN/m², the effective stress increase (Δp) is equal to 200 kN/m², and the preconsolidation pressure (p_c) is 400 kN/m². In order to determine the initial void ratio, 100 cm³ of initial soil is selected and indicates that there are 45 cm³ of voids. What is the settlement of the soil?

- A. 20 cm
- B. 25 cm
- C. 30 cm
- D. 35 cm

13. The following diagram shows a uniform soil deposit with information about each layer. Assume that the surcharge acting at the top of the soil is 630 lb/ft². If the effective pressure at point A is 3,800 lb/ft², what is the effective density of the second layer?

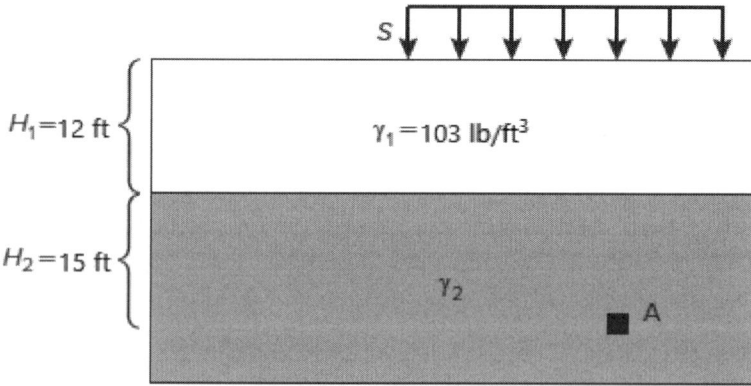

- A. 96 lb/ft³
- B. 117 lb/ft³
- C. 122 lb/ft³
- D. 129 lb/ft³

14. Find distance x from the right support B at which the shear force is zero. The beam is 20 ft long as shown. Assume that w = 10 kips/ft and b = 8 ft.

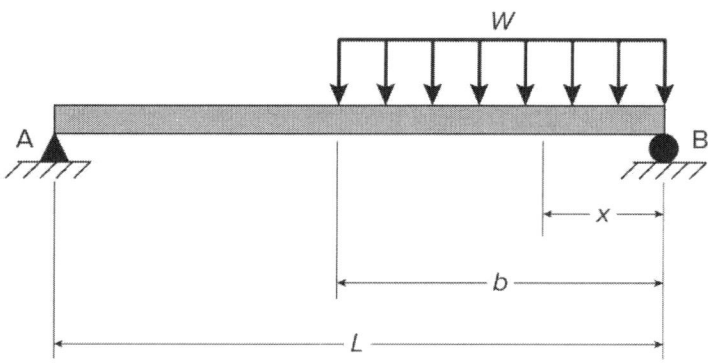

- A. 3.2 ft
- B. 5.6 ft
- C. 6.4 ft
- D. 7.9 ft

15. If an inclined partition wall on a beam has a distributed load of $w = (10 + 10x)$ kN/m, where x is the distance from the left support, determine the amount of shear (V) and moment (M) at point O shown below. Assume that R = 200 kN and L = 10 m.

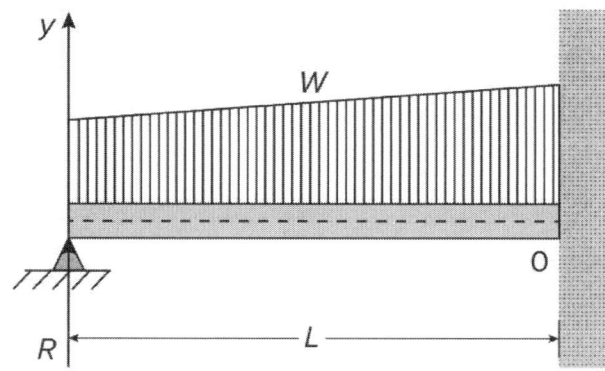

- A. V = 100 kN, M = 26,666 kN·m
- B. V = 250 kN, M = 4,333 kN·m
- C. V = 400 kN, M = 166.7 kN·m
- D. V = 500 kN, M = 3,333 kN·m

PRACTICE EXAM

16. Assume that $F = 800$ kips, $\theta = 30°$, $a = 6$ ft, $b = 3$ ft, the cross-sectional areas of parts A and B are 5 in² and 10 in² respectively, and $E = 29,000$ ksi. What is the elongation in the system below?

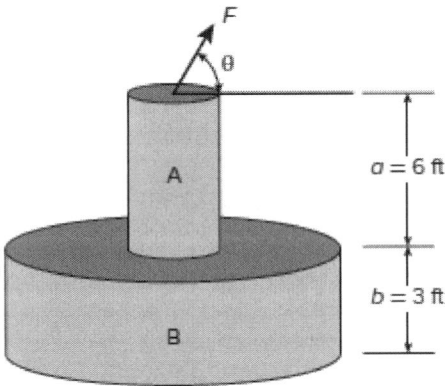

A. 0.25 in
B. 0.5 in
C. 1.75 in
D. 2.5 in

17. In the system below, what is the ratio of the lengths of cable 1 to cable 2 that keeps the rigid, weightless rod AB in a horizontal position? Assume that $E_1 = E_2$, $F = 800$ kN, $x = 1$ m, $A_1 = 2,600$ mm², $A_2 = 1,300$ mm², and $AB = L = 4$ m.

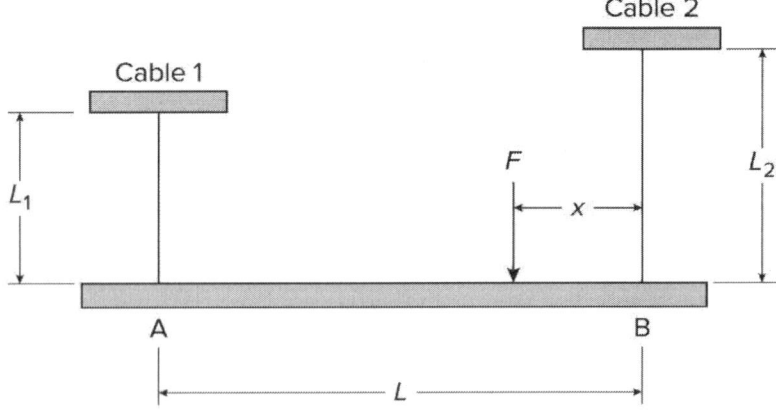

A. 0.25
B. 0.5
C. 1
D. 6

18. A 4-m cantilever beam is loaded with a distributed load of 1 kN/m. If the beam has a rectangular cross section that is 10 cm wide × 25 cm deep, what is the maximum shearing stress at the support?

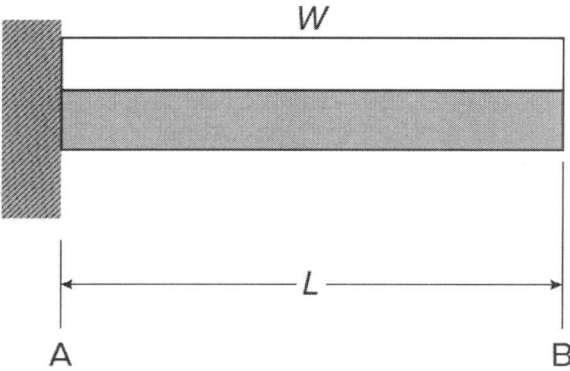

A. τ_{max} = 240 kPa
B. τ_{max} = 260 kPa
C. τ_{max} = 270 kPa
D. τ_{max} = 380 kPa

19. Two 18-in diameter pipes carry water into a pipe junction, and a single 30-in diameter pipe carries water out of the junction. The velocities in the two 18-in pipes are 2.5 ft/s and 3.8 ft/s. What is the velocity in the 30-in pipe?

A. 1.98 ft/s
B. 2.27 ft/s
C. 3.15 ft/s
D. 4.27 ft/s

20. A pipe system with a corrugated metal pipe (n = 0.024) and a slope of 0.02 ft/ft is being designed. The desired flow is 200 ft³/s. What is the smallest diameter pipe required to carry this flow?

A. 1.0 ft
B. 3.4 ft
C. 5.0 ft
D. 9.6 ft

PRACTICE EXAM

21. A pressurized water main is 400 ft long with a 24-in diameter, a Hazen-Williams friction factor C of 110, and a head loss of 2.4 ft. What is the average velocity in this pipe?

A. 4.12 ft/s
B. 5.91 ft/s
C. 7.43 ft/s
D. 9.12 ft/s

22. A sanitary sewer 15 inches in diameter carries an open channel flow. The downstream end of the pipe is open, and the water flow from the pipe forms a waterfall. At the end of the pipe, the depth is equal to half of the pipe diameter. Compute the discharge Q in the pipe.

A. 0.62 ft³/s
B. 1.23 ft³/s
C. 2.41 ft³/s
D. 24.1 ft³/s

23. Water is flowing through 55 ft of 8-in pipe at a rate of 250 gpm. The water also travels through three check valves ($K = 0.3$) and two gate valves ($K = 0.15$). The friction losses through the pipe are known to be 0.65 ft per 100 ft of pipe. What are the total losses in the system?

A. 0.21 ft
B. 0.32 ft
C. 0.41 ft
D. 0.80 ft

24. Runoff from a 3-acre site is to be drained by a channel. The time of concentration for this site is 40 min. The site has a runoff coefficient C of 0.2. Rainfall quantities to be used for design are 0.5 in for a 20-min storm, 0.7 in for a 40-min storm, and 0.9 in for a 60-min storm. For what discharge should this channel be designed?

A. 0.36 ft³/s
B. 0.63 ft³/s
C. 1.04 ft³/s
D. 4.30 ft³/s

25. The centerline of a four-lane roadway is a horizontal circular curve as shown below. It is known that the PC station is sta 8+60, the curve radius R is 2,480 ft, and the intersection angle I is 70°. What is most nearly the PT station?

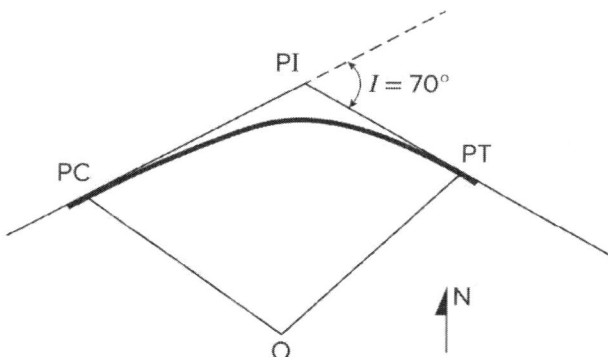

A. sta 38+15
B. sta 38+60
C. sta 39+89
D. sta 38+90

26. Assume that a speed-density study has resulted in the following calibrated relationship between space mean speed (v, mph) and density (k, veh/mi/ln): $v = 55 - 0.45k$. Determine the jam density and free-flow speed.

A. 0 veh/mi/ln, 55 mph
B. 120 veh/mi/ln, 50 mph
C. 122.2 veh/mi/ln, 0 mph
D. 122.2 veh/mi/ln, 55 mph

27. In horizontal curve formulas, what is tangent distance?

A. The distance from the PC to PI or from the PI to PT
B. The distance from the PI to the middle point of the curve
C. The distance along the line joining the PC and the PT
D. The distance from the middle point of the curve to the middle of the chord joining the PC and PT

PRACTICE EXAM

28. A horizontal curve is designed with an intersection angle of 13°21′55″. The degree of curvature (D) is 7.74°. Determine the tangent distance (T) to the nearest ft.

A. 87 ft
B. 118 ft
C. 542 ft
D. 740 ft

29. Increasing the water content in a concrete batch mix will result in (select all that apply):

A. decreased strength.
B. increased slump.
C. increased sulfate resistance.
D. increased strength.
E. decreased slump.

30. A soil fill sample has a weight of 62 lb, a total volume of 864 in³, and a water content of 15%. Determine the percent relative compaction of the sample if the maximum dry unit weight is 115 lb/ft³.

A. 93.8%
B. 95.2%
C. 102%
D. 127%

PRACTICE EXAM

31. Based on the boring log shown below, what is most nearly the buoyant (submerged) unit weight of soil at a depth of 15 ft? Assume that the soil below the water table is saturated.

DEPTH (ft)	N VALUE	UNIFIED SOIL CLASSIFICATION	DEPTH TO WATER (ft) 5.0	DEPTH TO WATER (ft) 5.0
			MOISTURE CONTENT (%)	DRY DENSITY (lb/ft³)
0	10	SM	10%	105
5				
6	7	CL	27%	90
10				
12				
15	8	CL	25%	95
19				
20	9	CL	21%	100

A. 33 lb/ft³
B. 56 lb/ft³
C. 95 lb/ft³
D. 119 lb/ft³

32. A soil has a void ratio of 0.7 and a water content of 22%. What is the total unit weight of the soil if the soil solids have a specific gravity of 2.7?

A. 102 lb/ft³
B. 112 lb/ft³
C. 121 lb/ft³
D. 210 lb/ft³

33. The construction of a new concrete pad has been commissioned to replace an existing one that has begun to deteriorate. The proposed slab is located in a coastal environment where there is a high presence of sulfates. In addition, the owner has mentioned that they will impose liquidated damages in the event the concrete does not meet expectations. Which type of cement should be specified for the batch mix design?

A. Type I
B. Type II
C. Type III
D. Type V

PRACTICE EXAM

34. What is the plastic section modulus ratio of hot-rolled steel sheet pile PZ 22 to PZ 35?

    A. 0.37
    B. 0.38
    C. 0.63
    D. 2.62

35. After the placement of a reinforced concrete wall in a commercial building, the owner suspects internal defects in the structural element due to the low slump of concrete (2 in) and wants to examine this possibility. No damage to the structure is allowed by the engineer. Which test method(s) can be used to identify the potential internal defects? Select all that apply.

    A. Impact-echo
    B. Sounding
    C. Infrared thermography
    D. GPR
    E. Impulse response

36. The Department of Transportation is looking to construct a proposed earthen road using borrow soil. The road will be shaped in the form of a trapezoidal cross section. The top width of the roadway will be 50 ft. The height of the roadway from existing grade is 4.5 ft. The roadway should be constructed with a minimum slope of 3:1 (H:V). The length of the new road will be 5 miles. A sample of the imported soil is tested and yields a shrinkage percentage of 20% and a swell percentage of 35%. What is the minimum bank volume required to construct the roadway?

    A. 279,400 yd^3
    B. 335,280 yd^3
    C. 349,250 yd^3
    D. 429,846 yd^3

PRACTICE EXAM

37. Given the earthwork summary at the following stations, determine the net quantity and resultant condition between sta 1+25 and sta 3+50 using the average end-area method.

STATION	FILL (ft²)	CUT (ft²)
1+25	110	20
3+50	50	230
4+50	30	42

A. 375 yd³ (cut)
B. 666.7 yd³ (fill)
C. 1,041.7 yd³ (cut)
D. 1,708.3 yd³ (fill)

38. A superintendent has been asked to compute the safety incidence rate for the last calendar year for insurance reporting purposes. From January to November, there have been a total of six serious injuries and three moderate illnesses due to exposures. From November to December, there was only one serious injury. There were 135 active employees during the reporting period, each working 56 hours per week for 50 weeks. The computed incidence rate (IR) is most nearly:

A. 3.70
B. 4.76
C. 5.29
D. 6.24

39. A recent geotechnical investigation revealed that the layered soil adjacent to the existing building can be classified as type C over type B according to OHSA regulation 1926 subpart P, "Excavations." An excavation is planned to upgrade subsurface utility structures. A 3-ft undisturbed perimeter is required from the face of the building, the total depth of excavation is 13 ft, and the top layer of soil is 6 ft deep. According to OSHA, the minimum horizontal distance from the face of the building to the toe of the slope is most nearly:

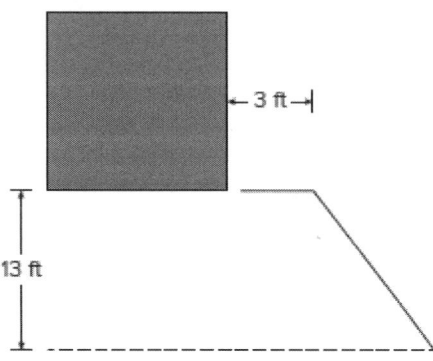

A. 16 ft
B. 19 ft
C. 19.5 ft
D. 22 ft

40. Based on the information provided in the figure below, the difference in elevation from the first benchmark (BM_1) to the second benchmark (BM_2) is most nearly:

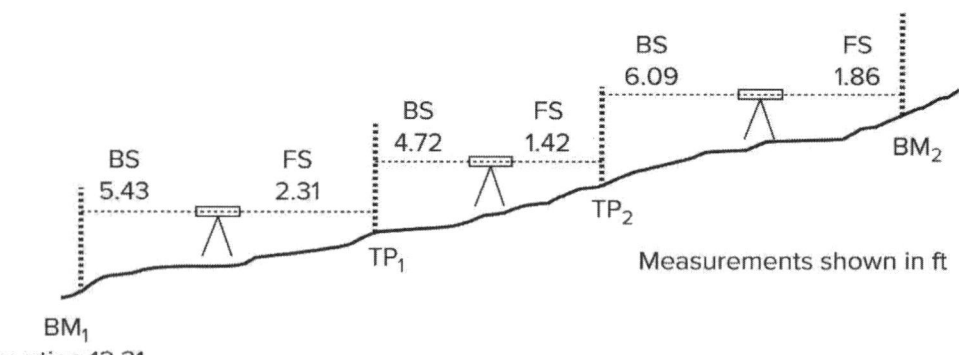

A. 1.66 ft
B. 10.65 ft
C. 22.96 ft
D. 28.55 ft

41. Determine the maximum shear force, bending moment, and axial force on the beam shown below:

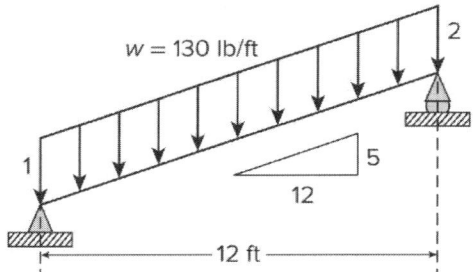

A. Maximum shear force = 780 lb; bending moment = 2,535 lb-ft; axial force = 325 lb
B. Maximum shear force = 780 lb; bending moment = 5,070 lb-ft; axial force = 325 lb
C. Maximum shear force = 845 lb; bending moment = 2,535 lb-ft; axial force = 0 lb
D. Maximum shear force = 845 lb; bending moment = 5,070 lb-ft; axial force = 0 lb

42. Based on the unfactored loads given below, determine the maximum vertical uniform load and maximum horizontal uniform load on the first floor of a concrete structure using the necessary load combinations according to load resistance factor design (LRFD).

The loads acting on the floor level per 1-ft strip are listed below.
Dead load = 110 lb/ft
Live load = 200 lb/ft
Wind load = 21 lb/ft
Horizontal earthquake load = 32 lb/ft

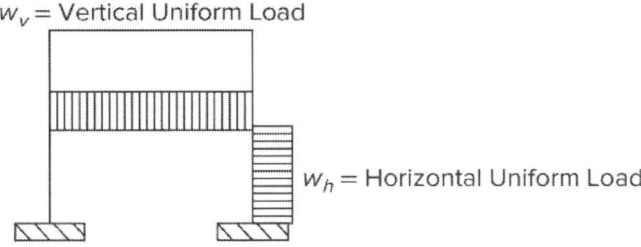

A. Maximum vertical load = 154 lb/ft; maximum horizontal load = 32 lb/ft
B. Maximum vertical load = 132 lb/ft; maximum horizontal load = 32 lb/ft
C. Maximum vertical load = 452 lb/ft; maximum horizontal load = 21 lb/ft
D. Maximum vertical load = 452 lb/ft; maximum horizontal load = 32 lb/ft

43. The building shown in the figure below is located in open terrain. Determine the maximum lateral load due to wind pressure if the wind velocity is 106 mph. Use K_d = 0.85, K_{zt} = 1 for flat ground.

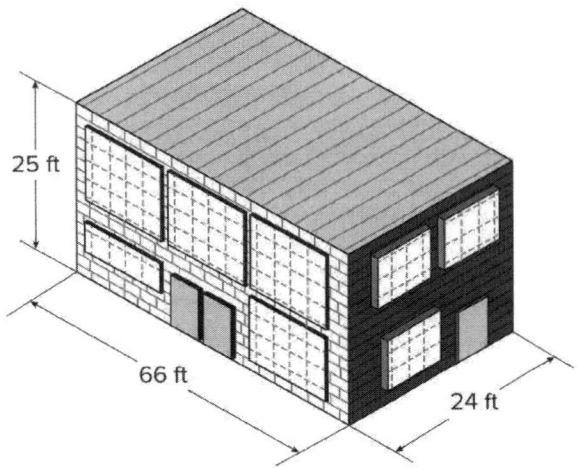

A. 20.98 lb/ft²
B. 22.98 lb/ft²
C. 24.98 lb/ft²
D. 26.98 lb/ft²

44. A crane is applying two moving loads 3 ft apart to a crane beam as shown in the figure. The maximum service load moment (ft-kips) carried by the beam is most nearly:

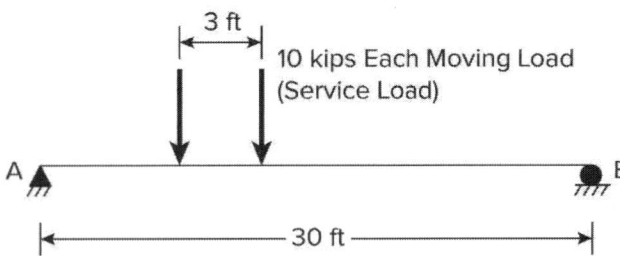

A. 100 kip-ft
B. 135 kip-ft
C. 150 kip-ft
D. 200 kip-ft

45. What is the uniform design roof snow load for the carport structure shown below? The structure is located in the center of a large parking lot with no nearby structures or foliage. Assume the terrain category for the structure's location is C, the ground snow load (p_g) of the location is 40 psf, and the carport deck has no heating components.

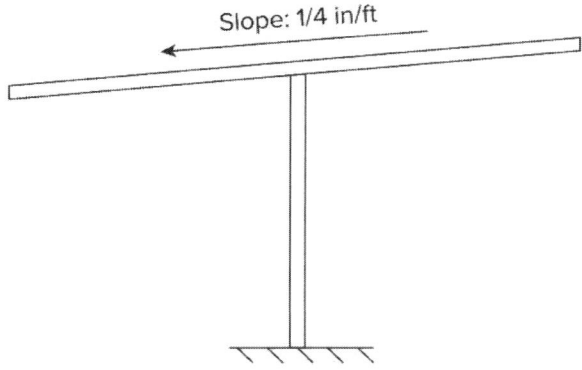

A. 22.68 psf
B. 27.72 psf
C. 28.02 psf
D. 30.24 psf

46. Determine tension and compression in the following member. $A_1 = 25$ in², $E_1 = 3,000$ ksi, $I_1 = 52$ in⁴, $L_1 = 3$ ft 0 in, $A_2 = 36$ in², $E_2 = 4,000$ ksi, $I_2 = 108$ in⁴, $L_2 = 5$ ft 0 in. The force (F) = 1 kip.

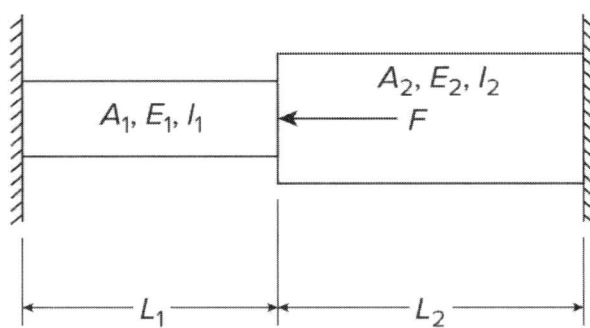

A. Compression = 0.46 kip; tension = 0.54 kip
B. Compression = 0.50 kip; tension = 0.50 kip
C. Compression = 0.37 kip; tension = 0.63 kip
D. Compression = 0.63 kip; tension = 0.37 kip

PRACTICE EXAM

47. In the structural layout shown below, assume there is a hinge at point C and the top of column connections act as pinned supports. The maximum moment at point B is _____ kip-ft.

Fill in the blank.

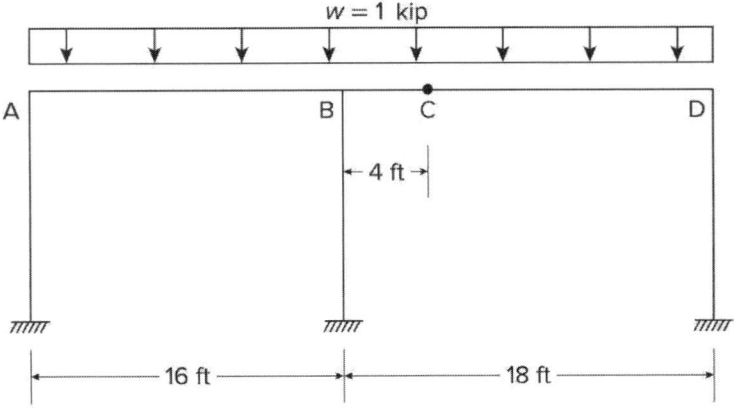

48. The hook shown below has a load $P = 2$ kips applied as indicated. Determine the maximum stress associated with location A-A.

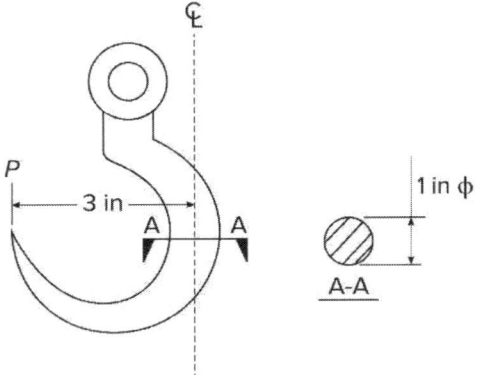

A. 3 ksi
B. 62 ksi
C. 64 ksi
D. 74 ksi

49. In the fall protection system shown below, determine the maximum force in the wire rope with the load shown and the rope sag, Δ = 3 ft 0 in.

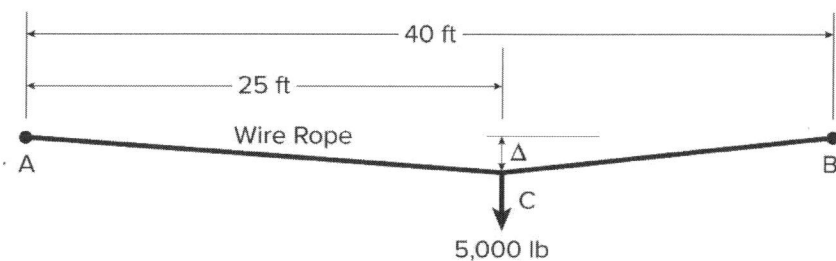

A. 1.9 kips
B. 3.2 kips
C. 15.7 kips
D. 16.0 kips

50. Determine the flexural strength demand of a W14 beam spaced at 6 ft on center on a roof with a surface dead load of 8 psf, which includes beam weight, surface live load of 20 psf, positive wind pressure of 25 psf, and uplift wind pressure of 75 psf. The roofing material is composed of metal roof deck and bituminous waterproof covering. The beam span is 30 ft and is simply supported. Use LRFD.

A. -46 kip-ft
B. -37 kip-ft
C. 37 kip-ft
D. 64 kip-ft

51. Determine the design moment at the base of the handrail post shown by LRFD. Disregard the self-weight of the handrail.

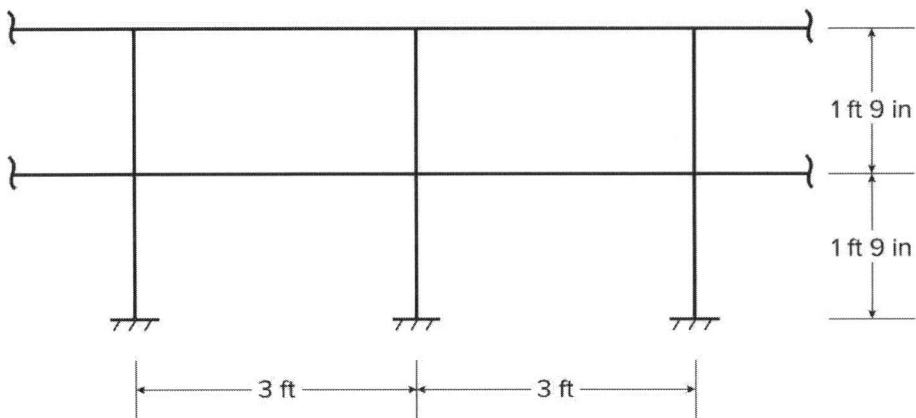

A. 525 ft-lb
B. 700 ft-lb
C. 840 ft-lb
D. 1,120 ft-lb

52. The podium slab shown supports condominiums. The total seismic weight at the podium slab = 1,150 kips. The slab is supported on concrete columns with the dimensions shown. Determine the maximum force in column C-2 in Y direction if C_s = 0.2 and column ends are considered fixed at base and pinned at slab. Ignore amplified torsion effects. Concrete compressive strength = 4,000 psi.

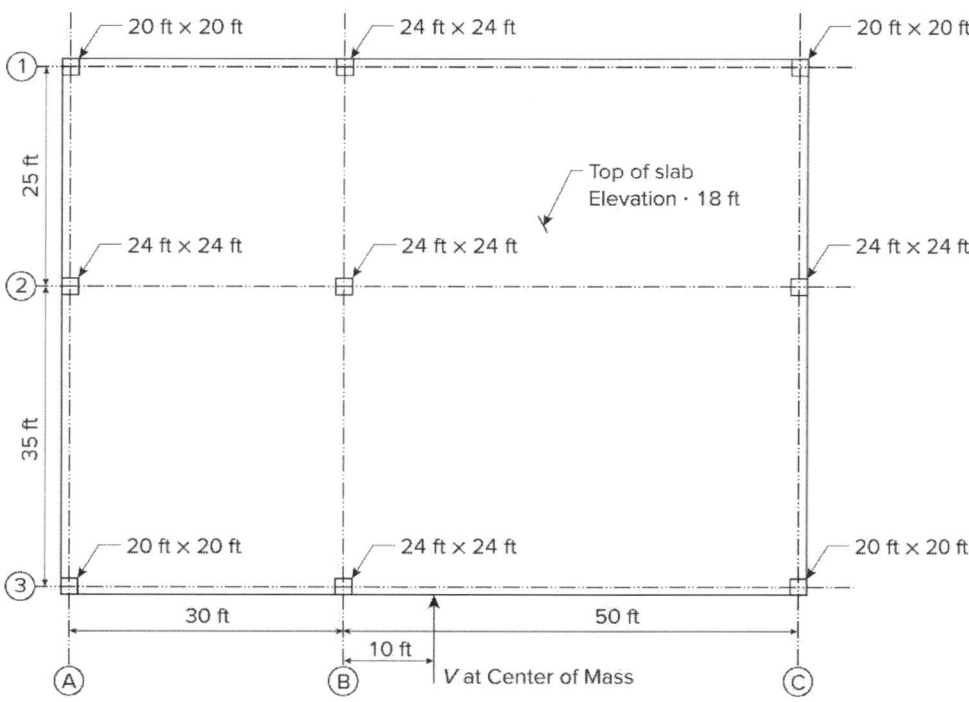

A. 25.6 kips
B. 33.2 kips
C. 34.1 kips
D. 41.7 kips

53. A three-story steel structure is located in Los Angeles, CA. The ground floor carries 700 kips, the first story carries 850 kips at a height of 15 ft, the second story carries 850 kips at a height of 12 ft, and the third story carries 700 kips at a height of 12 ft. Use the coefficient for the upper limit of the calculated period (C_u) and the mapped spectral response acceleration parameters S_s = 2.106 g, S_1 = 0.761 g to compute the base shear (V).
Criteria:
Site classification: C, steel special concentrically braced frames, occupancy category II

A. 123 kips
B. 200 kips
C. 307 kips
D. 562 kips

PRACTICE EXAM

54. According to ACI Code 318-14, what is the minimum concrete beam depth that resists cracking at the extreme fiber? The beam is under unfactored bending moment of 80 k-ft, and its width is 15 in. The concrete compressive strength is $f'_c = 4$ ksi.

A. 27 in
B. 27.5 in
C. 28 in
D. 28.5 in

55. In building structures, corrosion protection is not required for steel in which situations? Select all that apply.

A. Steel enclosed by building finish
B. Steel in contact with concrete
C. Steel in a dry climate
D. Steel in a paper-processing plant
E. Steel coated with contact-type fireproofing

56. Three 6 × 12-in standard cylinders were tested for compressive and tensile strength. A beam test was also performed on a 20 × 6 × 6-in beam using third-point loading. The following table shows the results of the tests.

BATCH NUMBER	SPLITTING TENSILE STRENGTH FAILURE LOAD (lb)	BEAM TEST FAILURE LOAD (lb)	COMPRESSIVE STRENGTH FAILURE LOAD (lb)
1	52,500	2,780	123,000
2	49,600	3,320	132,200
3	51,000	3,100	127,560

What is the ratio of tensile strength to compressive strength for batch 3?

A. 10%
B. 14%
C. 16%
D. 20%

PRACTICE EXAM

57. A steel anchor rod bracket is welded to a column face with two vertical fillet welds. The anchor rod bracket is 1 in thick and the eccentricity between the weld group and the anchor rod is 18 in. The factored load in the anchor rod is 100 kips acting vertically on the anchor rod bracket. Assuming the weld length is 24 in and E70 electrodes are used, what is the minimum weld size required using LRFD?

   A. 3/16 in
   B. 1/4 in
   C. 5/16 in
   D. 3/8 in

58. A bracing member is required to brace the truss below. The two factored lateral forces could occur on the structure. Using LRFD, find the smallest A36 3 1/2-in leg angle that will satisfy both loading conditions. The angle section is attached to a gusset plate at joints C and B. The load is placed at mid-width of the long leg of the angle at a distance of 0.28 in from the face of the leg.

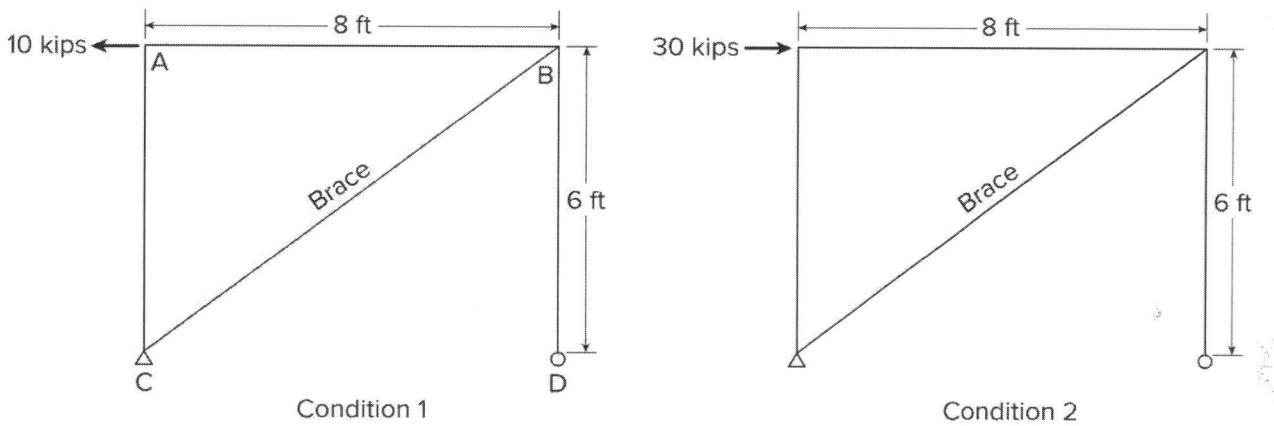

   A. L3 1/2 × 3 1/2 × 1/4
   B. L3 1/2 × 3 1/2 × 5/16
   C. L3 1/2 × 3 1/2 × 3/8
   D. L3 1/2 × 3 1/2 × 1/2

59. A 6 × 6-in reinforced concrete column has 4-#7 longitudinal bars and #3 ties spaced every 6 in. Assume grade 60 steel and a concrete compressive strength of 4,000 psi. What is the design compressive strength of the column?

   A. 143 kips
   B. 134 kips
   C. 241 kips
   D. 296 kips

60. A 3 × 3-ft square column is centered on a 10 × 10-ft square foundation. The foundation is 1.5 ft thick; the steel is grade 60, the longitudinal bars are 10- #10 in each way, and the clear cover is 3 in. The unfactored column loads of D = 50 kips and L = 85 kips are applied on the foundation. What is the flexure moment on the foundation?

    A. 100 kip-ft
    B. 116 kip-ft
    C. 137 kip-ft
    D. 154 kip-ft

61. Twelve piles are equally spaced in both directions in the foundation shown below. If forces of P = 300 kips, M_x = 500 kip-ft, and M_y = 350 kip-ft are applied to the center of the foundation, what is the maximum pile compression force?

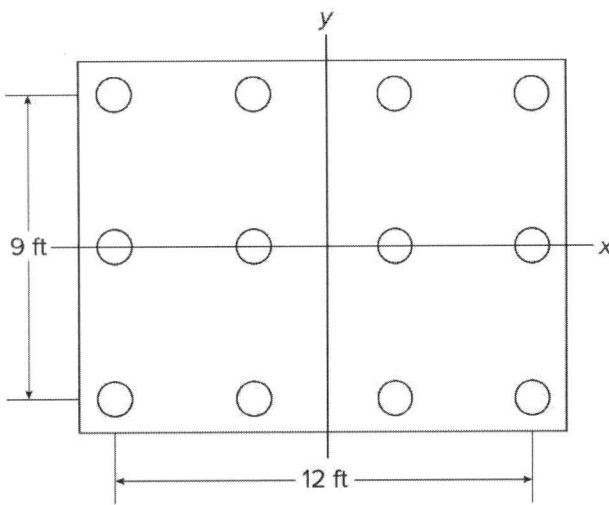

    A. 25.0 kips
    B. 32.2 kips
    C. 47.7 kips
    D. 49.4 kips

62. A 2 × 2-ft square pier is centered on a 7 × 7-ft pad foundation. The pad is 2 ft thick; the concrete compressive strength is 3.5 ksi, normal weight concrete, the longitudinal bars are #10, and the clear cover is 3 in. What is the design two-way shear strength of the foundation?

    A. 568.4 kips
    B. 642.7 kips
    C. 668.7 kips
    D. 734.2 kips

63. What is the allowable axial capacity for a 14 × 14 Mountain Hemlock No. 1, 25-ft tall column? The member is located indoors with a moisture content less than 19%. The column will carry dead and live loads.

   A. 80 kips
   B. 90 kips
   C. 100 kips
   D. 125 kips

64. A spread column footing is 8 × 14 ft and is under 81 kip-ft (in the long dimension) and 500 kips axial load. What is the maximum bearing pressure under the footing?

   A. 1.2 ksf
   B. 2.4 ksf
   C. 4.8 ksf
   D. 9.6 ksf

65. Given the foundation wall shown, at which point should the vertical reinforcement be placed to counteract the soil stresses imposed on the concrete wall? Circle the correct letter.

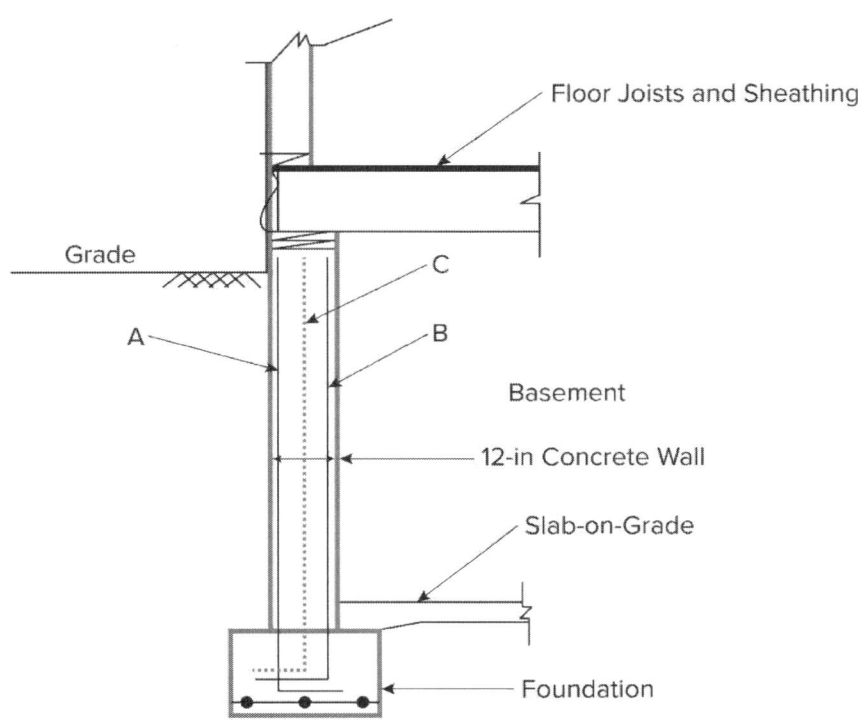

66. What is most nearly the capacity of a class A slip-critical bolt group consisting of six 1-in-diameter A325 bolts connecting a steel angle to the flange of a wide-flange beam that is subjected to the following loads: T_u = 50 kips and V_u = 45 kips?

    A. 65 kips
    B. 89 kips
    C. 104 kips
    D. 130 kips

67. For the wood stud size shown, determine the number of studs required to support a vertical load of 10 kips. Assume stud K = 1.0, the stud weak axis is completely braced, and ignore eccentricities and out of plane loads.

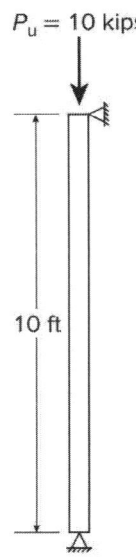

Given:
2 × 6 visually graded dimension lumber, Spruce Pine Fir #2
$C_D = C_M = C_t = C_i = 1.0$

    A. One stud
    B. Two studs
    C. Three studs
    D. Four studs

68. Given the 8-in fully grouted concrete masonry unit (CMU) wall shown, determine the spacing required for #4 horizontal steel shear reinforcing bars.

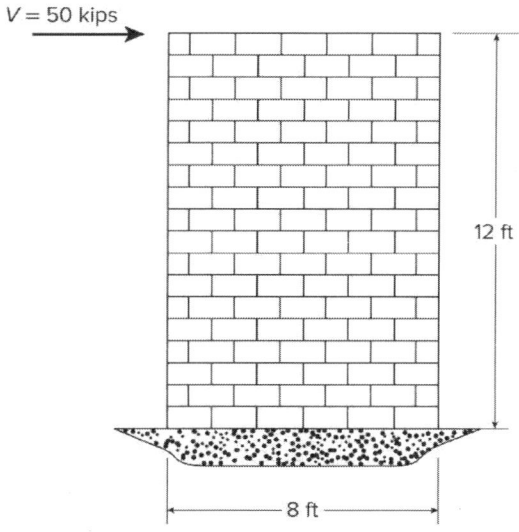

$f'_m = 1{,}500$ psi
$f_y = 60$ ksi
$V = $ ASD wind load

A. 8 in
B. 16 in
C. 24 in
D. 32 in

69. Given the material properties and sizes of the masonry and steel reinforcement shown below, determine the maximum allowable moment (ASD) the beam can withstand. Check bending only.

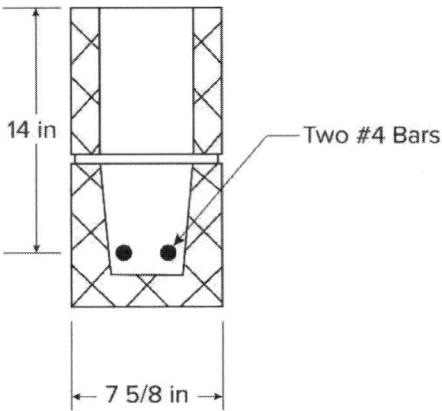

Masonry properties:
$E_s = 29,000$ ksi
$f_y = 60,000$ psi
$f'_m = 1,500$ psi
$F_s = 32,000$ psi

A. 9 kip-ft
B. 10 kip-ft
C. 12 kip-ft
D. 13 kip-ft

70. Assume that both ends of a steel column are pinned. What is the critical buckling load of a column, given the following data?

$E = 29,000$ ksi
$I_x = 100$ in⁴
$I_y = 60$ in⁴
$r_x = 2.3$ in
$r_y = 1.78$ in
$L = 10$ ft
$A = 18.9$ in²
$F_y = 50$ ksi

A. 678 kips
B. 774 kips
C. 1,044 kips
D. 1,985 kips

71. Determine the maximum bolt shear with the configuration shown using the elastic method.

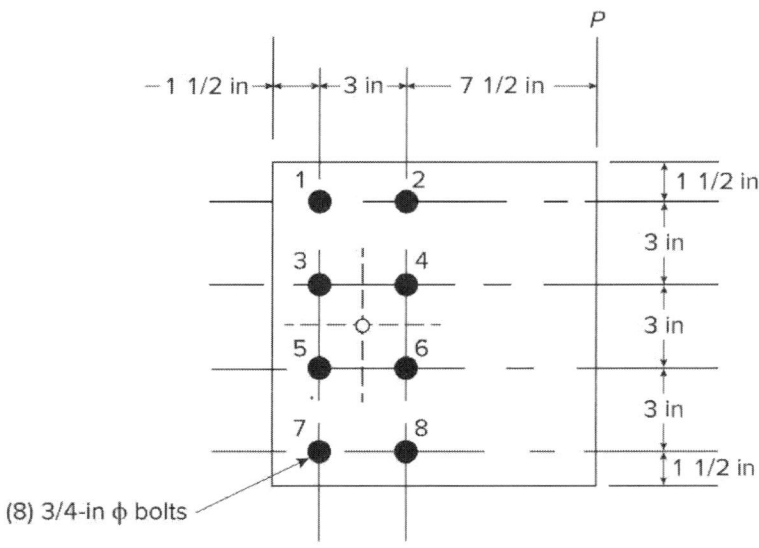

$P = 20$ kips

- A. 2.5 kips
- B. 6.1 kips
- C. 7.9 kips
- D. 9.0 kips

72. A flat bar that is 6.5 in wide, 5/8 in thick, and 2 ft long is welded through a longitudinal weld of 12 in. What is the effective net area for the welded member?

- A. 2.30 in²
- B. 3.53 in²
- C. 4.06 in²
- D. 5.03 in²

73. For the structure shown, all columns are W12 × 58 and all girders are W14 × 132. Assuming the columns buckle about their strong axis, what is the effective column length factor (K) of column AB?

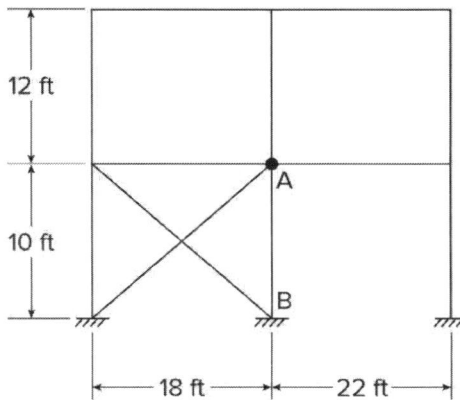

A. 0.65
B. 1.0
C. 1.25
D. 2.0

74. What is the approximate LRFD design axial load of a 60-ft long W14 × 132 column as shown below? (A36 steel is used.)

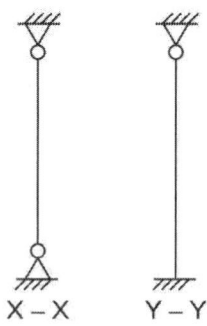

A. 9.65 kips
B. 374 kips
C. 1,380 kips
D. 9,650 kips

PRACTICE EXAM

75. A non-sway reinforced concrete column is under an axial dead load of 30 kips and a live load of 25 kips (unfactored) induced by transverse loads applied between supports. Critical buckling load is 210 kips. What is the magnification factor δ for the column?

- A. 1.0
- B. 1.9
- C. 2.8
- D. 3.5

76. A cross-laminated timber has a nominal char rate of 1.5 in/hr based on a 1-hr exposure. What is the effective char rate for 2.25 hrs of exposure (in/hr)?

- A. 1.5 in/hr
- B. 1.55 in/hr
- C. 1.67 in/hr
- D. 1.8 in/hr

77. What would be the deflection limit of a cantilevered section of roadway for a four-lane concrete bridge with a covered walkway and part of the exterior lane that are cantilevered beyond the exterior girder on both sides of the roadway?

- A. Span/800
- B. Span/240
- C. Span/360
- D. Span/375

78. A symmetric prefabricated building (8 ft wide × 40 ft long × 8 ft high) is to be lifted by a crane utilizing four cables as shown in the figure (at four corners). The cables are connected to the vertical hoisting cable at a point 12 ft directly above the center of gravity of the building. If the weight of the prefabricated building is 2 tons, the tension in the cables is most nearly:

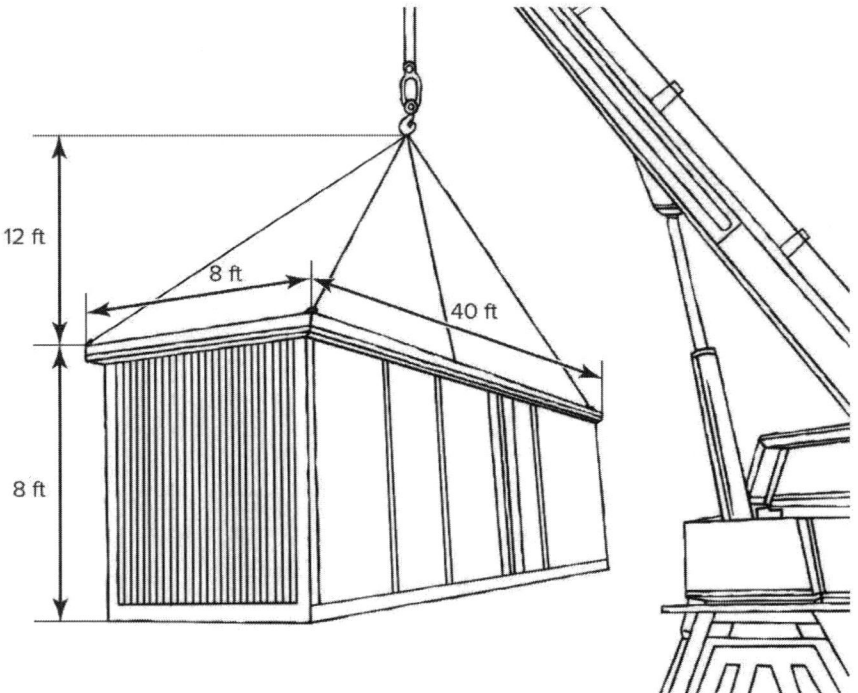

- A. 1,000 lb
- B. 1,971 lb
- C. 3,942 lb
- D. 4,000 lb

79. The prefabricated jacket shown below weighs 4,000 lb. During erection, it is lifted using two cables as shown. Find the working loads on the shackle if a safety factor of 2 is needed.

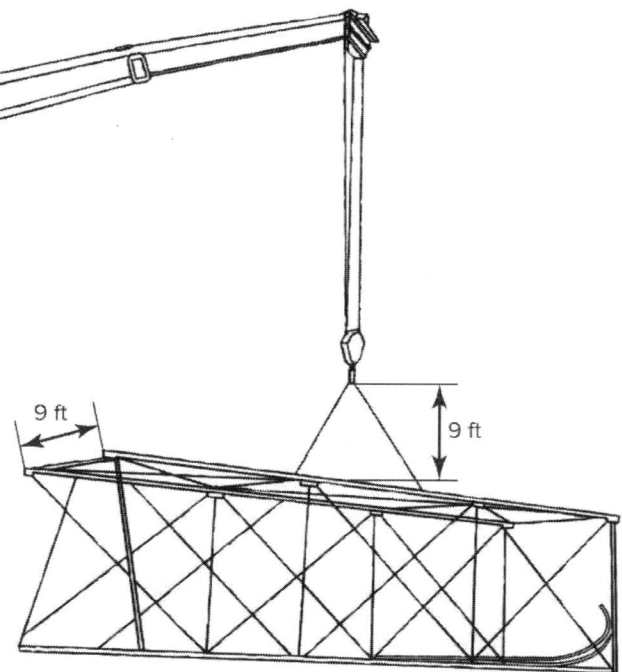

A. 1,788 lb
B. 2,000 lb
C. 4,000 lb
D. 4,472 lb

80. The portable ladder shown in the figure below is used to access a top surface. The minimum length of the extended rail (*a*) is 3 ft, and the horizontal distance of (*b*) is _____ ft.

Fill in the blank.

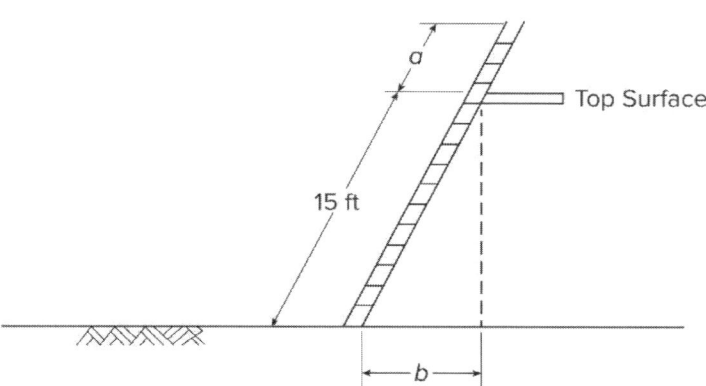

SOLUTIONS

QUICK SOLUTIONS REFERENCE

Question #	Answer
1.	C
2.	A
3.	355
4.	B
5.	51
6.	C
7.	C
8.	C
9.	D
10.	B
11.	A
12.	A
13.	D
14.	C
15.	C
16.	A
17.	D
18.	A
19.	B
20.	C
21.	B
22.	C
23.	C
24.	B
25.	D
26.	D
27.	A
28.	A
29.	A, B
30.	A
31.	B
32.	C
33.	D
34.	B
35.	A, E
36.	C
37.	A
38.	C
39.	B
40.	B

Question #	Answer
41.	A
42.	D
43.	B
44.	B
45.	D
46.	A
47.	36
48.	C
49.	D
50.	A
51.	D
52.	D
53.	D
54.	D
55.	A, B, E
56.	A
57.	C
58.	C
59.	B
60.	C
61.	C
62.	B
63.	B
64.	C
65.	point A
66.	B
67.	B
68.	B
69.	C
70.	A
71.	D
72.	B
73.	A
74.	B
75.	B
76.	B
77.	D
78.	B
79.	D
80.	3.75

SOLUTIONS

1. First, calculate the length of the channel wall:

$$\text{Hyp} = \left\{ \sqrt{(42 \text{ ft})^2 + \left[\frac{1}{4}(42 \text{ ft})\right]^2} \right\} = 43.29 \text{ ft}$$

Calculate the volume of concrete given the 7.5% waste:

$$V_{\text{Concrete}} = \left(\frac{8 \text{ in}}{12 \text{ in}}\right)\{[650 \text{ ft}][30 \text{ ft} + 2(43.29 \text{ ft})]\} = 50{,}518 \text{ ft}^3$$

$$+ \text{Waste} = 1.075(V_{\text{Concrete}}) = 54{,}306.9 \text{ ft}^3$$

Convert to cubic yards:

$$V_{\text{Concrete}} = (54{,}309.3 \text{ ft}^3)\left(\frac{1 \text{ yd}^3}{27 \text{ ft}^3}\right)$$

$$V_{\text{Concrete}} = 2{,}011.5 \text{ yd}^3$$

Reference: NCEES *PE Civil Reference Handbook* > Construction > Estimating Quantities and Costs
Answer: C

2. Perform the forward pass to determine the early dates (ES and EF). Then, perform the backward pass to determine the late dates (LS and LF). Once both the early and late dates are determined, compute the total float by employing the following equation: TF = LS − ES or LF − EF.

	ES	LS	EF	LF	FLOAT
1	0	0	3	3	0
2	3	7	13	17	4
3	3	3	9	9	0
4	9	9	17	17	0
5	17	17	21	21	0

Alternatively, recognize that after the forward pass, activity 5 is present on the critical path, and the total and free float of critical activities are always equal to zero. Therefore, the total float of task 5 is zero.

Reference: NCEES *PE Civil Reference Handbook* > Construction > Scheduling > Critical Path Method (CPM) Network Analysis
Answer: A

SOLUTIONS

3. Determine the cost of the entire building:

$$(100{,}000 \text{ ft}^2)\left(\frac{\$360}{90 \text{ ft}^2}\right) = \$400{,}000$$

Determine the budget remaining after the cost of the building and how many parking spots can be built with that amount:

$$\$2{,}000{,}000 - \$400{,}000 = \$1{,}600{,}000$$

$$\text{Number of parking spots} = \frac{\$1{,}600{,}000}{\$4{,}500/\text{spot}} = 355.55 = 355 \text{ spots}$$

Reference: NCEES *PE Civil Reference Handbook* > Construction > Estimating Quantities and Costs

Answer: 355

4. The following relationships are true for any activity:

TF = LF − EF

EF = ES + D

Substitute:
TF$_C$ = LF$_C$ − (ES$_C$ + D$_C$) = LF$_C$ − ES$_C$ − D$_C$

Reference: NCEES *PE Civil Reference Handbook* > Construction > Scheduling > Critical Path Method (CPM) Network Analysis

Answer: B

SOLUTIONS

5. Determine the minimum daily production rate to complete the scope (or the amount of material that must be hauled in a given day to meet the schedule of 15 days):

$$\frac{182{,}212 \text{ yd}^3/\text{job}}{15 \text{ days/job}} = 12{,}147.47 \text{ yd}^3/\text{day}$$

Determine how much material a single truck can haul in a given 8-hr day, recognizing that a single truck can transport 20 yd³ per trip (cycle) over the span of 40 min (0.667 hr):

$$(20 \text{ yd}^3/\text{cycle})\left(\frac{8 \text{ hr/day}}{0.667 \text{ hr/cycle}}\right) = 240 \text{ yd}^3/\text{day (per truck)}$$

Determine how many trucks are required to support a production of 12,147.47 yd³ if every truck can haul 240 yd³ per day:

$$\frac{12{,}147.47 \text{ yd}^3/\text{day}}{240 \text{ yd}^3/\text{day (per truck)}} = 50.61 \text{ trucks} \approx 51 \text{ trucks}$$

Reference: NCEES *PE Civil Reference Handbook* > Construction Operations and Methods > Production Rate for Loading and Hauling Earthwork

Answer: 51

SOLUTIONS

6. Moment of force (or torque) is equal to the force multiplied by the perpendicular distance to the point of inquiry:
$$M = Fd_\perp = 750 \text{ in} \cdot \text{lb}$$

Solve for the perpendicular distance assuming a maximum force of 50 lb can be applied:
$$\frac{M}{F} = d_\perp = \frac{750 \text{ in} \cdot \text{lb}}{50 \text{ lb}} = 15 \text{ in}$$

Recognize that the 15 in will be composed of 2 distances: the distance from the center of the bolt to the offset, and the leg of the triangle (with a hypotenuse of L).
15 in – 1.75 in = 13.25 in

13.25 is the perpendicular distance from the bolt head to the application of force, P.

Solve for distance L using the angle:
$$\frac{13.25 \text{ in}}{\cos 30°} = 15.3 \text{ in}$$

Reference: NCEES *PE Civil Reference Handbook* > Statics
Answer: C

SOLUTIONS

7. Recognize that the opposing force components are attributed from the wind and brace resistance, respectively.

The maximum wind load is a function of the tributary area, which is determined as:
A_{Trib} = (height of the wall)(center-to-center brace spacing)
A_{Trib} = 12 ft(7 ft) = 84 ft²

Solve for the component of the maximum brace load:
$F = \dfrac{F_x}{\cos \theta} = 4{,}158 \text{ lb}; \ F_x = 4{,}158 \text{ lb}(\cos 45°) = 2{,}940.15 \text{ lb}$

Solve for the sum of moments at point A to determine the opposing point load (P) generated from W_{Load} to F_x (the resultant of the wind load (P) is applied at the middle of the wall height):
$\sum M_A = \left(\dfrac{12}{2}P\right) - [(6 \text{ ft})(2{,}940.15 \text{ lb})] = 0; \ P = 2{,}940.15 \text{ lb}$

Solve for maximum W_{Load} given A_{Trib}:
$P = W_{Load} A_{Trib} = 2{,}940.15 \text{ lb}; \ \dfrac{2{,}940.15 \text{ lb}}{84 \text{ ft}^2} = 35 \text{ lb/ft}^2$

Reference: *NCEES PE Civil Reference Handbook* > Statics
Answer: C

8. The diagram shows that there are three layers of soil with unit weights of 16 kN/m³, 17 kN/m³, and 18 kN/m³.

The total stress at point A is calculated as:
$\sigma = \gamma_1 H_1 + \gamma_2 H_2 + \gamma_3 H_3$
$\sigma = (16 \text{ kN/m}^3)H_1 + (17 \text{ kN/m}^3)(3 \text{ m}) + (18 \text{ kN/m}^3)(2 \text{ m})$
$\sigma = (87 + 16H_1) \text{kPa}$

The pore water pressure is given as:
$u = \gamma_w (H_2 + H_3) = (9.8 \text{ kN/m}^3)(3 \text{ m} + 2 \text{ m}) = 49 \text{ kPa}$

Effective pressure at point A is calculated as:
$\sigma' = \sigma - u = (87 + 16H_1) \text{kPa} - 49 \text{ kPa} = 90 \text{ kPa}$

The above equation shows:
$H_1 = 3.25 \text{ m} \approx 3.3 \text{ m}$

Reference: NCEES *PE Civil Reference Handbook* > Geotechnical > Effective and Total Stresses
Answer: C

SOLUTIONS

9. It is known that:

Water content, $\omega = 100 \frac{W}{W_S}$

Specific gravity of soil solids, $G_S = \frac{W_S}{V_S \gamma_w}$

Void ratio, $e = \frac{V_V}{V_S}$

Degree of saturation, $S = 100 \frac{V_w}{V_V}$

Water volume, $V_W = \frac{W_W}{\gamma_w}$

Therefore:

$Se = 100 \left(\frac{V_w}{V_V}\right)\left(\frac{V_V}{V_S}\right) = 100 \frac{V_w}{V_S}$ and $G_S\omega = 100 \left(\frac{W_S}{V_S\gamma_w}\right)\left(\frac{W_w}{W_S}\right) = 100 \frac{V_w}{V_S}$

Which means that: $Se = G_S\omega$

The void ratio of the soil above the water level is calculated as:

$e = \frac{G_S\omega}{S} = \frac{2.7(20\%)}{0.5} = 1.08$

The unit weight of soil above the water level is calculated as:

$\gamma_A = \gamma_W \left[\frac{G_S + Se}{1+e}\right] = 9.8 \text{ kN/m}^3 \left[\frac{2.7 + 0.5(1.08)}{1+1.08}\right] \cong 15.27 \text{ kN/m}^3$

The void ratio of the soil below the water level is calculated as:

$e = \frac{G_S\omega}{S} = \frac{2.7(35\%)}{1.0} = 0.945$

The unit weight of soil below the water level is calculated as:

$\gamma_B = \gamma_W \left(\frac{G_S + e}{1+e}\right) = 9.8 \text{ kN/m}^3 \left(\frac{2.7 + 0.945}{1+0.945}\right) \cong 18.37 \text{ kN/m}^3$

Effective stress is equal to the total stress minus pore water pressure:
$\sigma' = \sigma - u$.

The total stress is calculated as: $\sigma = \gamma_A H_1 + \gamma_B H_2$
$\sigma = (15.27 \text{ kN/m}^3)(3 \text{ m}) + (18.37 \text{ kN/m}^3)(5 \text{ m}) = 137.66 \text{ kPa}$

The pore water pressure is calculated as:
$u = \gamma_W H_2 = (9.8 \text{ kN/m}^3)(5 \text{ m}) = 49 \text{ kPa}$

The effective stress is calculated as:
$\sigma' = \sigma - u = 137.66 \text{ kPa} - 49 \text{ kPa} = 88.66 \text{ kPa}$

Reference: NCEES *PE Civil Reference Handbook* > Effective and Total Stresses & Material Test Methods
Answer: D

SOLUTIONS

10. Though a soil friction angle is not given, it is related to the slope of the failure plane, which can be used directly to compute the Rankine active earth pressure coefficient:

$$K_a = \tan^2\left(45 - \frac{\phi}{2}\right) \rightarrow \alpha = 45 + \frac{\phi}{2} \rightarrow K_a$$
$$= \tan^2(90 - \alpha) = \tan^2(30) = 0.33$$

The horizontal stress at the base of the wall is:
$\sigma_h = K_a \gamma H$

$\sigma_h = (0.33)(115 \text{ lb/ft}^3)(15 \text{ ft})$

$\sigma_h = 569.25 \text{ lb/ft}^2$

The active lateral earth pressure resultant is then:
$R_a = \sigma_h H / 2$

$$R_a = \frac{(569.25 \text{ lb/ft}^2)(15 \text{ ft})}{2}$$

$R_a = 4{,}269.38 \text{ lb/ft} \approx 4{,}270 \text{ lb/ft}$

Reference: NCEES *PE Civil Reference Handbook* > Lateral Earth Pressures > Rankine Earth Coefficients

Answer: B

SOLUTIONS

11. Settlement for soil in the range of virgin compression is calculated as:
$$\Delta H = \frac{H_0}{1+e_0}\left[C_C \log \frac{p_0 + \Delta p}{p_0}\right]$$

In the equation above, all values are given in the question except the coefficient of consolidation (C_C) and the initial void ratio (e_0). Settlement is also referenced in units of feet.
First determine the initial void ratio:
$$e = \frac{n}{1-n} = \frac{0.5}{1-0.5} = 1$$

Based on the equation above, the C_C is calculated as:
$$\Delta H = \frac{H_0}{1+e_0}\left[C_C \log \frac{p_0 + \Delta p}{p_0}\right] = \frac{20 \text{ ft}}{1+1}\left[C_C \log \frac{(3{,}000 \text{ lb/ft}^2 + 2{,}000 \text{ lb/ft}^2)}{3{,}000 \text{ lb/ft}^2}\right]$$
$$= 12 \text{ in} = 1 \text{ ft}$$

Therefore:
$$C_C \approx 0.45$$

References: NCEES *PE Civil Reference Handbook* > Geotechnical > Material Test Methods > Weight-Volume Relationships
NCEES *PE Civil Reference Handbook* > Geotechnical > Consolidation
Answer A

12. Settlement for soil is calculated as:
$$\Delta H = \frac{H_0}{1+e_0}\left[C_R \log \frac{p_C}{p_0} + C_C \log \frac{p_0 + \Delta p}{p_C}\right]$$

The equation above shows everything except for the initial void ratio.
The void ratio is calculated as:
$$e = \frac{V_V}{V_S} = \frac{V_V}{V - V_V} = \frac{45 \text{ cm}^3}{100 \text{ cm}^3 - 45 \text{ cm}^3} = 0.8$$

The settlement of the soil is calculated as:
$$\Delta H = \frac{5 \text{ m}}{1+0.8}\left[(0.1)\log\frac{400 \text{ kN/m}^2}{300 \text{ kN/m}^2} + (0.6)\log\frac{500 \text{ kN/m}^2}{400 \text{ kN/m}^2}\right] = 19.6 \text{ cm}$$

Reference: NCEES *PE Civil Reference Handbook* > Geotechnical > Consolidation
Answer: A

SOLUTIONS

13. The diagram shows that there are two layers of soil. For the first layer, the unit weight is equal to:
$\gamma_1 = 103 \text{ lb/ft}^3$

For the second layer, γ_2, the unit weight is unknown.
The effective stress at point A is calculated as:
$\sigma = S + \gamma_1 H_1 + \gamma_2 H_2$
$\sigma = 630 \text{ lb/ft}^2 + (103 \text{ lb/ft}^3)(12 \text{ ft}) + \gamma_2(15 \text{ ft}) = 3{,}800 \text{ lb/ft}^2$
$\sigma = 1{,}866 \text{ lb/ft}^2 + \gamma_2(15 \text{ ft}) = 3{,}800 \text{ lb/ft}^2$

The above equation shows:
$\Gamma_2 = 128.9 \text{ lb/ft}^3$

Reference: NCEES *PE Civil Reference Handbook* > Geotechnical > Effective and Total Stresses
Answer: D

14. Take moment about B to the find the reaction at A (the distributed load is converted to a point load acting at the center of b):

$$\sum M_B = 0$$

$$R_A(L) - w(b)\left(\frac{b}{2}\right) = 0$$

$$R_A = \frac{w(b)\left(\frac{b}{2}\right)}{L} = \frac{(10 \text{ kpf})(8 \text{ ft})\left(\frac{8 \text{ ft}}{2}\right)}{20 \text{ ft}} = 16 \text{ kips}$$

$$\sum F_y = 0$$

$$R_A + R_B - w(b) = 0$$

$$R_B = w(b) - R_A = 10 \text{ kpf } (8 \text{ ft}) - 16 \text{ kips} = 64 \text{ kips}$$

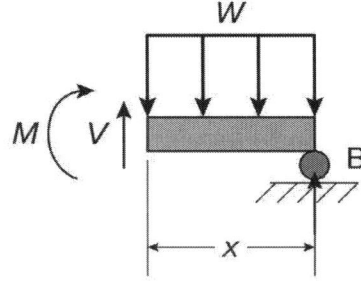

Cut a section at x from support B as seen in the beam free-body diagram above. The distance x from the right support with zero shear force is calculated as:

$$\sum F_V = 0$$

$$R_B - w(x) + V = 0$$

Set V equal to 0.

$$R_B - w(x) = 0$$

$$64 \text{ kips} - 10\frac{\text{kips}}{\text{ft}(x)} = 0$$

Therefore:

$$x = \frac{64 \text{ kips}}{10 \text{ kpf}}$$

$$x = 6.4 \text{ ft}$$

Reference: NCEES *PE Civil Reference Handbook* > Statics
Answer: C

SOLUTIONS

15. Based on geometry, the distributed load is converted to a concentrated load (W):

$W = (10 + 10x)$ kN/m from 0 to 10

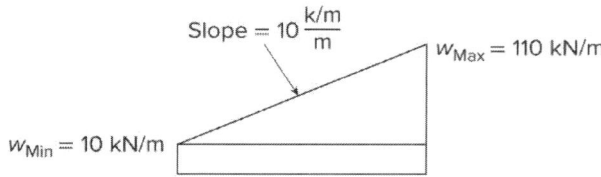

The minimum intensity at the left support ($x = 0$ m):

$w_{Min} = 10 \text{ kN/m} + 10 \dfrac{\text{kN/m}}{\text{m}} (0 \text{ m}) = 10 \text{ kN/m}$

The maximum intensity at point O (the right support, $x = 10$ m):

$w_{Max} = 10 \text{ kN/m} + 10 \dfrac{\text{kN/m}}{\text{m}} (10 \text{ m}) = 110 \text{ kN/m}$

Find the resultant of the rectangular distributed load:

$W_1 = w_{Min} L = 10 \dfrac{\text{kN}}{\text{m}} (10 \text{ m}) = 100 \text{ kN}$

Find the resultant of the triangular distributed load:

$W_2 = \dfrac{(w_{Max} - w_{Min})L}{2} = \dfrac{(110 \text{ kN/m} - 10 \text{ kN/m})10 \text{ m}}{2} = 500 \text{ kN}$

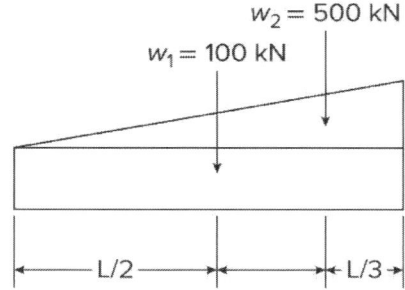

The shear force is calculated as

$\Sigma F = 0$

$R - W_1 - W_2 + V = 0$

$V = -R + W_1 + W_2 = -200 \text{ kN} + 100 \text{ kN} + 500 \text{ kN}$

$V = 400 \text{ kN}$

SOLUTIONS

The moment is calculated as:
$\Sigma M_O = 0$ (assume CW moment is positive)
$$R(10 \text{ m}) - W_1\left(\frac{10 \text{ m}}{2}\right) - W_2\left(\frac{10 \text{ m}}{3}\right) + M = 0$$

$$M = (-200 \text{ kN})(10 \text{ m}) + (100 \text{ kN})(5 \text{ m}) + (500 \text{ kN})\left(\frac{10 \text{ m}}{3}\right)$$

$M = -2{,}000 \text{ kN} \cdot \text{m} + 500 \text{ kN} \cdot \text{m} + 1{,}666.7 \text{ kN} \cdot \text{m}$

$M = 166.7 \text{ kN} \cdot \text{m}$

Reference: NCEES *PE Civil Reference Handbook* > Statics
Answer: C

16. Draw the following free-body diagram:

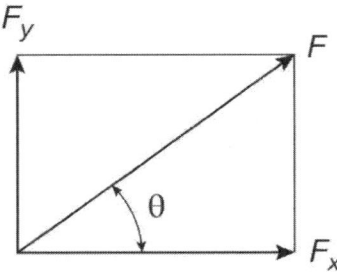

The force component that causes elongation is F_y:
$\Sigma F_y = 0$

$F_y = F \sin \theta = 800 \text{ kips } (\sin 30°) = 400 \text{ kips}$

Elongation is calculated as:
$$\delta_{\text{Total}} = \delta_A + \delta_B = \frac{F_y L_1}{E A_A} + \frac{F_y L_2}{E A_B}$$

$$\delta_{\text{Total}} = \frac{(400 \text{ kips})(6 \text{ ft})(12 \text{ in/ft})}{(5 \text{ in}^2)(29{,}000 \text{ kips/in}^2)} + \frac{(400 \text{ kips})(3 \text{ ft})(12 \text{ in/ft})}{(10 \text{ in}^2)(29{,}000 \text{ kips/in}^2)} = 0.2472 \text{ in}$$

$\delta_{\text{Total}} \approx 0.25 \text{ in}$

Reference: NCEES *PE Civil Reference Handbook* > Mechanics of Materials > Uniaxial Stress-Strain
Answer: A

SOLUTIONS

17. The free-body diagram shown below is used to calculate F_1 and F_2:

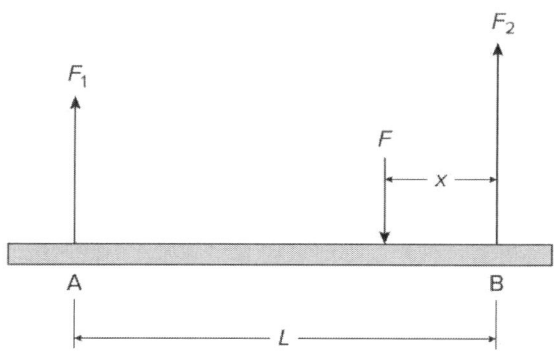

$\Sigma M_B = 0$

$F_1(4 \text{ m}) - (800 \text{ kN})(1 \text{ m}) = 0$

$F_1 = 200 \text{ kN}$

$\Sigma F_y = 0$

$F = F_1 + F_2$

$800 \text{ kN} = 200 \text{ kN} + F_2$

$F_2 = 800 \text{ kN} - 200 \text{ kN} = 600 \text{ kN}$

To keep the bar balanced, the deflection of cable 1 is equal to the deflection of cable 2, which is expressed as:

$\delta_1 = \delta_2$

$\delta_1 = \dfrac{F_1 L_1}{E_1 A_1}$

$\delta_2 = \dfrac{F_2 L_2}{E_2 A_2}$

$\dfrac{F_1 L_1}{E_1 A_1} = \dfrac{F_2 L_2}{E_2 A_2}$

Solve for the ratio L_1/L_2 with $E_1 = E_2$:

$\dfrac{L_1}{L_2} = \dfrac{F_2 E_1 A_1}{F_1 E_2 A_2} = \dfrac{(600 \text{ kN})(2{,}600 \text{ mm}^2)}{(200 \text{ kN})(1{,}300 \text{ mm}^2)} = 6$

Reference: NCEES *PE Civil Reference Handbook* > Mechanics of Materials > Uniaxial Stress-Strain

Answer: D

SOLUTIONS

18. The maximum shear force is at the fixed support and equal to the reaction at support A. The reaction at support A is calculated as:

$R_A = wL$
$R_A = (1 \text{ kN/m})(4 \text{ m}) = 4 \text{ kN}$
$R_A = V_{Max} = 4 \text{ kN} = 4{,}000 \text{ N}$

The following is necessary to calculate the shear stress:

A^* = area above the plane where the shear acts

y = distance from the neutral axis to the centroid of A^*

$Q = A^*y$

The maximum Q is at $h/2$ (mid depth of cross section):

$A^* = \left(\dfrac{bh}{2}\right) = \dfrac{(10 \text{ cm})(25 \text{ cm})}{2} = 125 \text{ cm}^2 = 0.0125 \text{ m}^2$

$y = \dfrac{h}{4} = \dfrac{25 \text{ cm}}{4} = 6.25 \text{ cm} = 0.0625 \text{ m}$

$Q_{Max} = (0.0125 \text{ m}^2)(0.0625 \text{ m}) = 0.0007812 \text{ m}^3$

The moment of inertia (I) is calculated as:

$\dfrac{bh^3}{12} = \left(\dfrac{1}{12}\right)(10 \text{ cm})(25 \text{ cm})^3 = 13{,}020.8 \text{ cm}^4 = 0.0001302 \text{ m}^4$

Maximum shearing stress is calculated as:

$\tau_{Max} = \dfrac{V_{Max} Q_{Max}}{Ib} = \dfrac{(4{,}000 \text{ N})(0.0007812 \text{ m}^3)}{(0.0001302 \text{ m}^4)(0.1 \text{ m})}$

$\tau_{Max} = 240{,}000 \text{ N/m}^2 = 240 \text{ kPa}$

Reference: NCEES *PE Civil Reference Handbook* > Mechanics of Materials > Beams

Answer: A

SOLUTIONS

19. Pipes 1 and 2 are the 18-in pipes, and pipe 3 is the 30-in pipe. The continuity equation states:

$V_1 A_1 + V_2 A_2 = V_3 A_3$

$A_1 = A_2 = \dfrac{\pi D^2}{4} = \dfrac{3.14 \left(\dfrac{18}{12}\right)^2}{4} = 1.77 \text{ ft}^2$

$A_3 = \dfrac{3.14 \left(\dfrac{30}{12}\right)^2}{4} = 4.91 \text{ ft}^2$

$V_1 = 2.5 \text{ ft/s}$

$V_2 = 3.8 \text{ ft/s}$

Solve the continuity equation for V_3:

$V_3 = \dfrac{(V_1 A_1 + V_2 A_2)}{A_3}$

$V_3 = 2.27 \text{ ft/s}$

Reference: NCEES *PE Civil Reference Handbook* > Hydraulics > Principles of One-Dimensional Fluid Flow

Answer: B

SOLUTIONS

20. Manning's equation: $Q = \dfrac{KAR_H^{\frac{2}{3}}S^{\frac{1}{2}}}{n}$

Where:

K = 1.486 for USCS units

For circular pipes flowing full, R_H is $D/4$:

$$Q = \dfrac{1.486\left(\pi\dfrac{D^2}{4}\right)\left(\dfrac{D^{\frac{2}{3}}}{4}\right)\left(S^{\frac{1}{2}}\right)}{n} \rightarrow 2.16\dfrac{nQ}{\sqrt{S}} = D^{\frac{8}{3}}$$

Solve for D:

$$D = 1.335\left(\dfrac{nQ}{\sqrt{S}}\right)^{3/8}$$

Q = 200 ft³/s

n = 0.024

S = 0.02 ft/ft

$$D = 1.335\left(\dfrac{0.024(200)}{(0.02)^{1/2}}\right)^{3/8} = 5 \text{ ft}$$

Reference: NCEES *PE Civil Reference Handbook* > Open Channel Flow > Manning's Equation

Answer: C

SOLUTIONS

21. The Hazen-Williams equation for the average velocity is:

$V = k_1 C R_H^{0.63} S^{0.54}$

In the above equation, $k_1 = 1.318$ for USCS units, R_H = hydraulic radius, C = Hazen Williams coefficient, and S = friction slope (head loss per unit pipe length).

$k_1 = 1.318$

$R_H = \dfrac{\frac{24}{12}}{4} = 0.5 \text{ ft}$

$C = 110$

$S = \dfrac{2.4}{400} = 0.006 \text{ ft/ft}$

$V = 1.318(110)(0.5)^{0.63}(0.006)^{0.54}$

$V = 1.318(110)(0.646)(0.063)$

$V = 5.91 \text{ ft/s}$

Reference: NCEES *PE Civil Reference Handbook* > Hydraulics> Closed Conduit Flow and Pumps > Hazen-Williams Equation
Answer: B

SOLUTIONS

22. The depth of flow at the downstream end of a channel where the water descends in a waterfall is the critical depth. Under these conditions, the following expression applies:

$$\frac{Q^2}{g} = \frac{A^3}{T}$$

Where:
A = cross-sectional area
T = top width at the downstream end of the channel (or pipe in this case)

For a full pipe:

$$A_F = \pi \left(\frac{\frac{15}{12}}{2}\right)^2 = 1.23 \text{ ft}^2$$

If the pipe is half full, then the area is half the full value, or $A = 0.61$ ft².
For a half-full circular pipe, the top width T is equal to the pipe diameter D of 1.25 ft.

Solve the above equation for Q:

$$Q^2 = \frac{gA^3}{T}$$

$$Q^2 = \frac{(32.2 \text{ ft/s}^2)(0.61 \text{ ft}^2)^3}{1.25 \text{ ft}} = 5.85$$

$Q = 2.41$ ft³/s

Reference: NCEES *PE Civil Reference Handbook* > Hydraulics > Open Channel Flow > Specific Energy
Answer: C

SOLUTIONS

23. Determine the friction losses in the pipe:

Friction losses in pipe $= \dfrac{0.65 \text{ ft}}{100 \text{ ft}}(55 \text{ ft}) = 0.358 \text{ ft}$

Determine the velocity:

$Q = 250 \text{ gpm}(0.134 \text{ ft}^3/\text{gal})\left(\dfrac{1 \text{ min}}{60 \text{ s}}\right) = 0.56 \text{ ft}^3/\text{s}$

Area of the pipe $= \pi\left(\dfrac{4 \text{ in}}{12 \text{ in/ft}}\right)^2 = 0.35 \text{ ft}^2$

$V = \dfrac{Q}{A}$

$V = \dfrac{0.56 \text{ ft}^3/\text{s}}{0.35 \text{ ft}^2} = 1.6 \text{ ft/s}$

Determine the velocity head (h_V):

$h_V = \dfrac{V^2}{2g} = \dfrac{(1.6 \text{ ft/s})^2}{2(32.2 \text{ ft/s}^2)} = 0.04 \text{ ft}$

Determine the friction loss from the velocity head:

$h_M = h_V K$

K values are given in the question statement.

Friction loss in three check valves = (0.04)(0.3)(3) = 0.036 ft

Friction loss in two gate valves = (0.04)(0.15)(2) = 0.012 ft

Sum friction loss:
Total losses = 0.358 ft + 0.036 ft + 0.012 ft = 0.41 ft

Reference: NCEES *PE Civil Reference Handbook* > Hydraulics > Principles of One-Dimensional Fluid Flow

Answer: C

SOLUTIONS

24. Apply the rational formula:

$Q = CiA$

Q = discharge (ft³/s)

C = runoff coefficient for the watershed

i = rainfall intensity (in/hr)

A = watershed area (acres)

When applying the rational method, rain falling over a time period equal to the time of concentration of the watershed should be used. In this case, the time of concentration is given as 40 min, or 0.67 hr, and the corresponding rainfall amount is 0.7 in.

Rainfall intensity:
$$i = \frac{0.7 \text{ in}}{(40 \text{ min})\left(\frac{1 \text{ hr}}{60 \text{ min}}\right)} = 1.04 \text{ in/hr}$$

Solve for the discharge:
$Q = CiA$
$Q = (0.2)(1.04)(3) = 0.63$ ft³/s

Reference: NCEES *PE Civil Reference Handbook* > Hydrology > Runoff Analysis
Answer: B

25. The length of the curve from PC to PT is:
$$L = \frac{RI\pi}{180°} = \frac{(2{,}480 \text{ ft})(70°)\pi}{180°} \cong 3{,}029.89 \text{ ft}$$

The station of PT is:
sta PT = sta PC+L = (sta 8+60) + 3,029.89 ft
sta PT = (sta 8+60) + (sta 30+29.89) = sta 38+89.89

Reference: NCEES *PE Civil Reference Handbook* > Transportation > Horizontal Design > Basic Curve Elements
Answer: D

SOLUTIONS

26. The jam density occurs when the speed is 0 mph, or:
$v = 55 - 0.4k = 0$

$k = \dfrac{55}{0.45} = 122.2$ veh/mi/ln

The free-flow speed occurs when the density is 0 veh/mi/ln, or:
$v = 55 - 0.45(0) = 55$ mph

Reference: NCEES *PE Civil Reference Handbook* > Transportation > Traffic Engineering > Uninterrupted Flow
Answer: D

27. In horizontal curve formulas, tangent distance is the distance from the PC to PI or from the PI to PT (Answer A is correct).

Answer B is incorrect. The distance from the PI to the middle point of the curve is external distance.

Answer C is incorrect. The distance along the line joining the PC and the PT is the long chord.

Answer D is incorrect. The distance from the middle point of the curve to the middle of the chord joining the PC and PT is the middle ordinate.

Reference: NCEES *PE Civil Reference Handbook* > Transportation > Horizontal Design > Basic Curve Elements
Answer: A

SOLUTIONS

28. Use the arc definition:

$$R = \frac{5{,}729.58}{D}$$

$$R = \frac{5{,}729.58}{7.74°} = 740.3 \text{ ft}$$

Find the tangent length formula:

$$T = R \tan\left(\frac{\Delta}{2}\right)$$

$$\Delta = 13°21'55''$$

Convert to decimal:

$$13° + \left(\frac{21}{60}\right) + \left(\frac{55}{3{,}600}\right) = 13.35°$$

$$T = (740.3 \text{ ft})\left[\tan\left(\frac{13.35°}{2}\right)\right] = 86.6 \text{ ft} \approx 87 \text{ ft}$$

Reference: NCEES *PE Civil Reference Handbook* > Transportation > Horizontal Design > Basic Curve Elements
Answer: A

29. Increasing the water content in concrete will increase the workability (slump), making it easier for placement as expected since this will increase the cement paste content. However, because not all water is consumed during the cement hydration, the final concrete product will have a higher porosity as a result of the additional water. Therefore, the increased water content will compromise the strength in the process.

Reference: NCEES *PE Civil Reference Handbook* > Material Quality Control and Production > Material Properties and Testing
Answer: A, B

SOLUTIONS

30. The relative compaction cannot exceed 100%, so answers C and D can be eliminated. To determine the relative compaction, the dry unit weight of the fill sample must be determined first. Then, compare it to the maximum dry unit weight.

$$\gamma_d = \frac{\frac{W_T}{V_T}}{1 + \frac{w\%}{100\%}}$$

$$\gamma_d = \frac{\frac{62 \text{ lb}}{864 \text{ in}^3 \left(\frac{1 \text{ ft}}{12 \text{ in}}\right)^3}}{1 + \frac{15\%}{100\%}} = \frac{\frac{62 \text{ lb}}{0.5 \text{ ft}^3}}{1.15}$$

$\gamma_d = 107.8 \text{ lb/ft}^3$

The relative compaction is the dry unit weight of the sample divided by the maximum unit weight:

$$100\% \left(\frac{\gamma_d}{\gamma_{d,\max}}\right) = 100\% \left(\frac{107.8 \text{ lb/ft}^3}{115 \text{ lb/ft}^3}\right) = 93.8\%$$

Reference: NCEES *PE Civil Reference Handbook* > Geotechnical > Weight-Volume Relationships
Answer: A

31. The soil below the water table is saturated, and $\gamma = \gamma_{Sat} = \gamma_{Dry}(1 + w)$.

$\gamma_{Sat} = 95 \text{ lb/ft}^3 (1 + 0.25) = 118.8 \text{ lb/ft}^3$

$\gamma_b = \gamma_{Sat} - \gamma_w$

$\gamma_b = 118.8 \text{ lb/ft}^3 - 62.4 \text{ lb/ft}^3 = 56.4 \text{ lb/ft}^3$

Reference: NCEES *PE Civil Reference Handbook* > Geotechnical > Weight-Volume Relationships
Answer: B

SOLUTIONS

32. Assume that the volume of soil solids (V_S) is 1 ft³. The void ratio (e) is 0.7. The volume of water and air ($V_V = eV_S$) is 0.7 ft³.

The weight of the soil solids is:
$W_S = V_S G \gamma_W = (1 \text{ ft}^3)(2.7)(62.4 \text{ lb/ft}^3) = 168.48 \text{ lbf}$

The weight of water is:
$W_W = wW_S = 22\%(168.48 \text{ lb}) = 37.0656 \text{ lb}$

The weight of water and soil solids is:
$W = W_S + W_W = 168.48 \text{ lb} + 37.0656 \text{ lb} = 205.55 \text{ lb}$

Plot the known information in the phase diagram.

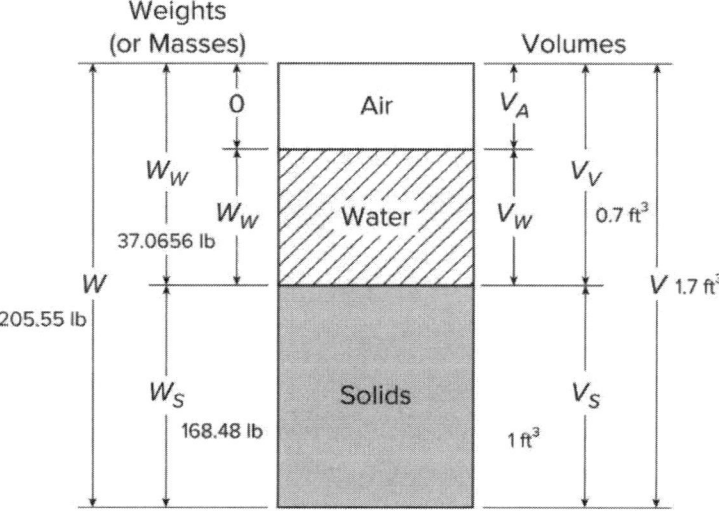

The total unit weight of the soil is:
$$\gamma = \frac{W}{V} = \frac{205.55 \text{ lb}}{1.7 \text{ ft}^3} \cong 120.9 \text{ lb/ft}^3$$

Reference: NCEES *PE Civil Reference Handbook* > Geotechnical > Weight-Volume Relationships

Answer: C

33. Type V cement is sulfate-resisting and is specified when there is extensive exposure to sulfates. This includes water with high alkali content and seawater.

Answer: D

SOLUTIONS

34. Use the "Properties of NZ and PZ Steel Sheet Piles" table in chapter 4, section 4.2.2 of the NCEES *PE Civil Reference Handbook*:

Plastic section modulus of PZ 22 = 21.79 in³/ft.

Plastic section modulus of PZ 35 = 57.17 in³/ft.

Therefore, the ratio is:

$$\frac{21.79}{57.17} = 0.38$$

Reference: NCEES *PE Civil Reference Handbook* > Steel Design > PZ Hot-Rolled Steel Sheet Pile Properties
Answer: B

35. According to the "Nondestructive Test Methods for Concrete" table in chapter 2, section 2.5.3.2 of the NCEES *PE Civil Reference Handbook*, internal defects can be identified by impact-echo, ultrasonic echo, and impulse response. Therefore, A and E are correct.

Reference: NCEES *PE Civil Reference Handbook* > Material Quality Control and Production > Concrete Maturity and Early Strength Evaluations
Answer: A, E

36. Recognize that the constructed condition is in compacted measure. Calculate the volume (compacted measure) of the roadway: $V = AL$

The cross section is a trapezoid with a top width of 50 ft. The height is 4.5 ft, and the slope is 3:1 (H:V).

Therefore, the bottom width is 50 ft + 2[3(4.5 ft)] = 77 ft.

$$A = \left[\left(\frac{50 \text{ ft} + 77 \text{ ft}}{2}\right)(4.5 \text{ ft})\right] = 285.75 \text{ ft}^2$$

Calculate volume:

$$V = (285.75 \text{ ft}^2)(5 \text{ miles})\left(\frac{5{,}280 \text{ ft}}{1 \text{ mile}}\right)\left(\frac{1 \text{ yd}^3}{27 \text{ ft}^3}\right) = 279{,}400 \text{ yd}^3$$

Convert compacted volume into bank volume:

$V_{\text{Compacted}} = (1 - S_H)V_{\text{Bank}}$
$279{,}400 \text{ yd}^3 = 0.80 V_{\text{Bank}}$
$V_{\text{Bank}} = 349{,}250 \text{ yd}^3$

Reference: NCEES *PE Civil Reference Handbook* > Construction > Earthwork Construction and Layout > Excavation and Embankment
Answer: C

SOLUTIONS

37. Between sta 1+25 and sta 3+50, the length is 225 ft. The volume of cut and fill are calculated as:

$$V = \left(\frac{A_1 + A_2}{2}\right)L$$

$$V_{Cut} = \left(\frac{20 \text{ ft}^2 + 230 \text{ ft}^2}{2}\right)\left(\frac{225 \text{ ft}}{27}\right) = 1,041.67 \text{ yd}^3$$

$$V_{Fill} = \left(\frac{110 \text{ ft}^2 + 50 \text{ ft}^2}{2}\right)\left(\frac{225 \text{ ft}}{27}\right) = 666.67 \text{ yd}^3$$

$$V_{Net} = V_{Cut} - V_{Fill} = 1,041.67 \text{ yd}^3 - 666.67 \text{ yd}^3 = 375 \text{ yd}^3 \text{ (cut)}$$

Reference: NCEES *PE Civil Reference Handbook* > Construction > Earthwork Construction and Layout > Earthwork Volumes
Answer: A

38. N = number of injuries, illnesses, and fatalities = 6 + 3 + 1 = 10

T = total hours worked by all employees during the period in question
T = 135[56(50)] = 378,000 hours

$$IR = \frac{200,000N}{T} = \frac{10(200,000)}{378,000} = 5.29$$

Reference: NCEES *PE Civil Reference Handbook* > Health and Safety > Safety Management and Statistics
Answer: C

39. Type C over type B soil requires slopes (H:V) of (1.5:1) and (1:1), respectively.

Type C = 1.5(6 ft) = 9 ft
Type B = 1(7 ft) = 7 ft

Total horizontal distance = undisturbed perimeter + horizontal distance associated with type C layer + horizontal distance associated with type B layer

Total horizontal distance = 3 ft + 9 ft + 7 ft = 19 ft

Reference: NCEES *PE Civil Reference Handbook* > Trench and Construction Safety > Slope Configurations: Excavations in Layered Soils
Answer: B

SOLUTIONS

40. To compute the difference in elevation from (2) points, compute the difference in the sum of both the Backsights and Foresights between the same (2) points, respectively.

Elevation of BM_1 + BS = HI → 12.31 ft + 5.43 ft = 17.74 ft
Elevation of TP_1 = HI − FS = 17.74 ft − 2.31 ft = 15.43 ft
Elevation of TP_1 + BS = HI → 15.43 ft + 4.72 ft = 20.15 ft
Elevation of TP_2 = HI − FS = 20.15 ft − 1.42 ft = 18.73 ft
Elevation of TP_2 + BS = HI → 18.73 ft + 6.09 ft = 24.82 ft
Elevation of BM_2 = HI − FS = 24.82 ft − 1.86 ft = 22.96 ft
Difference in elevation of BM_2 and BM_1 = 22.96 ft − 12.31 ft = 10.65 ft

The calculations are summarized in the table below:

POINT	BS (+)	HI	FS (−)	ELEVATION
BM_1	5.43	17.74		12.31
TP_1	4.72	20.15	2.31	15.43
TP_2	6.09	24.82	1.42	18.73
BM_2			1.86	22.96
	+16.24		−5.59	+10.65

Reference: NCEES *PE Civil Reference Handbook* > Construction > Earthwork Construction and Layout > Site Layout and Control
Answer: B

SOLUTIONS

41. In general, shear force and bending moment cannot be computed when the load is not perpendicular to the member length. Therefore, the uniform distributed load of 130 lb/ft is analyzed into two components. The vertical component is perpendicular to the member length, and the horizontal component is parallel to the member length.

$$w_v = 130 \text{ lb/ft} \left(\frac{12}{13}\right) = 120 \text{ lb/ft}$$

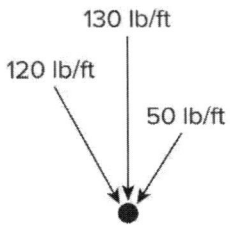

$$w_h = (130 \text{ lb/ft}) \left(\frac{5}{13}\right) = 50 \text{ lb/ft}$$

Shear force calculation:
The vertical reactions perpendicular to member length are:

$\sum M_1 = 0$

$$(13 \text{ ft})(R_{2v}) - \frac{(120 \text{ lb/ft})(13 \text{ ft})^2}{2} = 0$$

$R_{2v} = 780 \text{ lb}$

$\sum F'_y = 0$ perpendicular to the member

$R_{1v} + R_{2v} - w_v L = 0$

$R_{1v} + 780 \text{ lb} - (120 \text{ lb/ft})(13 \text{ ft}) = 0$

$R_{1v} = 780 \text{ lb}$

Bending moment calculation:

$$M = \frac{(R_{1v} L/2)}{2}$$

$$= \frac{(780 \text{ lb})(13 \text{ ft}/2)}{2} = 2{,}535 \text{ lb-ft}$$

or $M = \dfrac{w_v L^2}{8}$

$$= \frac{(120 \text{ lb/ft})(13 \text{ ft})^2}{8} = 2{,}535 \text{ lb-ft}$$

SOLUTIONS

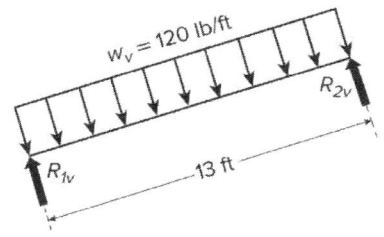
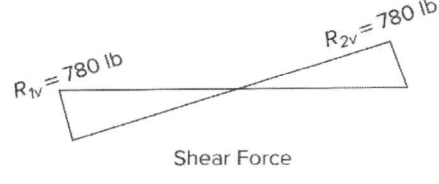
Shear Force
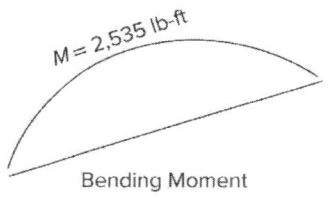
Bending Moment

Axial force calculation:
At support 2, R_2 can be computed from R_{2v}.

$R_{2v} = R_2 \left(\dfrac{12}{13}\right), \quad R_2 = (780 \text{ lb})\left(\dfrac{13}{12}\right) = 845 \text{ lb}$

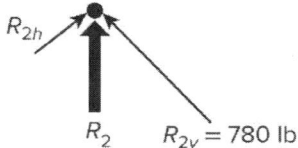

R_{2h} can be computed from R_2.

$R_{2h} = R_2 \left(\dfrac{5}{13}\right) = (780 \text{ lb})\left(\dfrac{5}{13}\right) = 325 \text{ lb}$

The horizontal reactions along the member length.

$\sum F'_x = 0$ horizontal reaction parallel to the member length

$R_{1h} + R_{2h} - w_h L = 0$

$R_{1h} = 50 \text{ lb/ft}\,(13 \text{ ft}) - 325 \text{ lb} = 325 \text{ lb}$

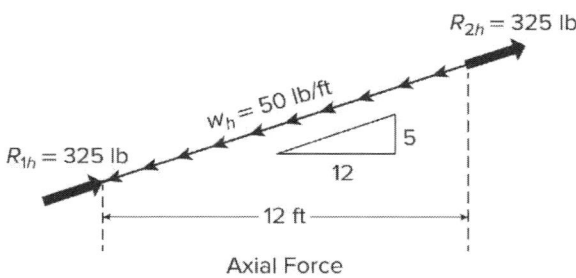
Axial Force

Reference: NCEES *PE Civil Reference Handbook* > Mechanics of Materials > Beams
Answer: A

SOLUTIONS

42. When the structural elements are designed according to LRFD, the basic combinations of the factored loads are presented in the following combinations:

1. $1.4D$
2. $1.2D + 1.6L + 0.5(L_r \text{ or } S \text{ or } R)$
3. $1.2D + 1.6(L_r \text{ or } S \text{ or } R) + (L \text{ or } 0.5W)$
4. $1.2D + 1.0W + L + 0.5(L_r \text{ or } S \text{ or } R)$
5. $1.2D + 1.0E + L + 0.2S$
6. $0.9D + 1.0W$
7. $0.9D + 1.0E$

Load on first floor:

LOAD COMBINATION	VERTICAL UNIFORM LOAD, W_v	HORIZONTAL UNIFORM LOAD, W_h
1) $w = 1.4(110 \text{ lb/ft}) =$	154 lb/ft	
2) $w = 1.2(110 \text{ lb/ft}) + 1.6(200 \text{ lb/ft}) =$	452 lb/ft	
3) $w = 1.2(110 \text{ lb/ft}) + (200 \text{ lb/ft}) =$	332 lb/ft	
4) $w = 1.2(110 \text{ lb/ft}) + (200 \text{ lb/ft}) + 1.0(21 \text{ lb/ft}) =$	332 lb/ft	21 lb/ft
5) $w = 1.2(110 \text{ lb/ft}) + 1.0(32 \text{ lb/ft}) + 200 \text{ lb/ft} =$	332 lb/ft	32 lb/ft
6) $w = 0.9(110 \text{ lb/ft}) + 1.0(21 \text{ lb/ft}) =$	99 lb/ft	21 lb/ft
7) $w = 0.9(110 \text{ lb/ft}) + 1.0(32 \text{ lb/ft}) =$	99 lb/ft	32 lb/ft

Maximum vertical load is 452 lb/ft from load combination 2, and maximum horizontal load is 32 lb/ft from load combinations 5 and 7.

Reference: ASCE/SEI Standard 7-10: *Minimum Design Loads for Buildings and Other Structures* > Section 2.3.2

Answer: D

SOLUTIONS

43. The lateral load on the given building is based on the wind velocity and geographic location.

The value of K_z is obtained from ASCE/SEI 7-10 Table 27.3-1 and Equation 27.3-1.

The average height of the structure is z = 25 ft from the figure in the problem statement and under open terrain (exposure C), K_z = 0.94 as shown in ASCE/SEI 7-10 section 26.7.2.

K_{zt} = 1.0

Wind velocity, V = 106 mph

by using the wind load pressure equation $q_z = 0.00256\, K_z K_{zt} K_d V^2$

$q_z = 0.00256(0.94)(1.0)(0.85)(106\text{ mph})^2 = 22.98 \text{ lb/ft}^2$

References: ASCE/SEI Standard 7-10: *Minimum Design Loads for Buildings and Other Structures* > Table 27.3-1 and Equation 27.3-1

Answer: B

44. Since a = 3 ft < 0.586 L = 0.586 (30 ft) = 17.58 ft, the maximum service moment is given as:

$$M = \frac{P\left(L - \frac{a}{2}\right)^2}{2L} = \frac{(10 \text{ kips})\left(30 \text{ ft} - \frac{3 \text{ ft}}{2}\right)^2}{(2)(30 \text{ ft})} \cong 135.4 \text{ kip} - \text{ft}$$

References: AISC *Steel* Manual 14th Edition > Table 3-23, case 44

Answer: B

SOLUTIONS

45. Snow load design is discussed in ASCE/SEI 7-10 chapter 7. The flat roof snow load must be calculated per Equation 7.3-1, which is then used to find the sloped roof snow load per Equation 7.4-1.

Equation 7.3-1:

$$p_f = 0.7 C_e C_t I_s p_g$$

Where:

C_e = exposure factor per Table 7.2
  = 0.9 (terrain category C given, structure fully exposed)

C_t = thermal factor per Table 7.3
  = 1.2 (Structure is unheated and open air)

I_s = importance factor per ASCE/SEI 7-10, Table 1.5-2 based on risk category from ASCE/SEI 7-10, Table 1.5-1
  = 1.0 (structure is considered risk category II)

Thus:
p_f = (0.7)(0.9)(1.2)(1.0)(40) = 30.24 psf

Equation 7.4-1:

$$p_s = C_s p_f$$

Where:

C_s = roof slope factor (per ASCE/SEI 7-10, Figure 7-2c; roof is unheated and can be considered cold)

C_s = 1.0 (roof slope given as 1/4 in:12 in, or 1.2°)

p_s = 1.0(30.24 psf) = 30.24 psf

Reference: ASCE/SEI Standard 7-10: *Minimum Design Loads for Buildings and Other Structures* > Chapter 7

Answer: D

SOLUTIONS

46. Due to the continuity of the member, a point at which there is a change in the section and material properties would move the same distance, whether we view it from right of section 2 or from left of section 1. It is easy to notice that section 1 is in compression, and section 2 in tension. The reaction on the left (R_1) and the reaction on the right (R_2) represent compression and tension forces, respectively, which must be calculated. We know their sum is equal to F, so:

$R_1 + R_2 = F \Rightarrow R_2 = F - R_1$

Knowing that the elongation of 2 is equal to the shortening of 1, we obtain:

$$\frac{R_1 L_1}{E_1 A_1} = \frac{R_2 L_2}{E_2 A_2} \Rightarrow \frac{R_1 L_1}{E_1 A_1} = \frac{(F - R_1) L_2}{E_2 A_2}$$

$$\Rightarrow \frac{R_1 L_1}{E_1 A_1} = \frac{F L_2}{E_2 A_2} - \frac{R_1 L_2}{E_2 A_2}$$

$$\Rightarrow \frac{R_1 L_1}{E_1 A_1} + \frac{R_1 L_2}{E_2 A_2} = \frac{F L_2}{E_2 A_2}$$

$$\Rightarrow R_1 \left(\frac{L_1}{E_1 A_1} + \frac{L_2}{E_2 A_2} \right) = \frac{F L_2}{E_2 A_2}$$

$$R_1 = \frac{\frac{F L_2}{E_2 A_2}}{\frac{L_1}{E_1 A_1} + \frac{L_2}{E_2 A_2}} = \frac{\frac{(1 \text{ kip} \times 5 \text{ ft} \times 12 \text{ in/ft})}{(4{,}000 \text{ ksi} \times 36 \text{ in}^2)}}{\left(\frac{3 \text{ ft} \times 12 \text{ in/ft}}{3{,}000 \text{ ksi} \times 25 \text{ in}^2} \right) + \left(\frac{5 \text{ ft} \times 12 \text{ in/ft}}{4{,}000 \text{ ksi} \times 36 \text{ in}^2} \right)} = 0.46 \text{ kip}$$

And

$R_2 = F - R_1 = 1 \text{ kip} - 0.46 \text{ kip} = 0.54 \text{ kip}$

Reference: NCEES *PE Civil Reference Handbook* > Mechanics of Materials > Definitions
Answer: A

SOLUTIONS

47. Determine shear at point C:

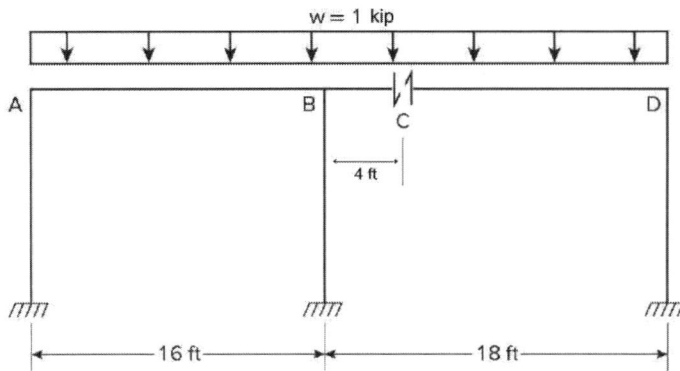

$$P \text{ at } C = \frac{wL_{CD}}{2} = \frac{(1 \text{ klf})(18 \text{ ft} - 4 \text{ ft})}{2} = 7 \text{ kips}$$

From statics, derive reactions at A and B.

$$\Sigma M_A = R_B(16 \text{ ft}) - P(20 \text{ ft}) - \frac{(1 \text{ klf})(20 \text{ ft})^2}{2} = 0$$

$R_B = 21.25$ kips

$\Sigma F_y = R_A + R_B - P - (1 \text{ klf})(20 \text{ ft})$

$R_A = 5.75$ kips

Create shear diagram with reactions and loadings:

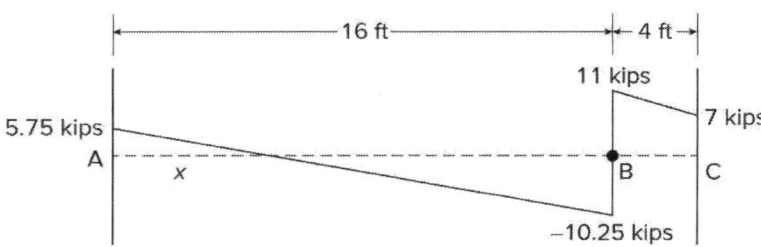

Calculate the area under shear diagram to right of point B for moment at B.

$$M_B = \frac{(11 \text{ kips} - 7 \text{ kips})(4 \text{ ft})}{2} + (7 \text{ kips})(4 \text{ ft}) = 36 \text{ kip} - \text{ft}$$

Also, you can calculate the area under shear diagram to the left of point B to compute the moment at B.

Distance from A to zero shear:

$$x = \frac{V}{w} = \frac{5.75 \text{ kips}}{1 \text{ kip/ft}} = 5.75 \text{ ft}$$

$$M_B = \frac{(5.75 \text{ kips})(5.75 \text{ ft})}{2} + \frac{(-10.25 \text{ kips})(16 \text{ ft} - 5.75 \text{ ft})}{2} = -36 \text{ kip} - \text{ft}$$

Reference: NCEES *PE Civil Reference Handbook* > Structural Analysis > Moment, Shear, and Deflection Diagrams

Answer: 36

SOLUTIONS

48. Determine the area and section modulus.

$$A = \frac{\pi d^2}{4} = 0.785 \text{ in}$$

From *Steel Manual*, Table 17-27:

$$S = \frac{\pi d^3}{32} = 0.098 \text{ in}^3$$

$$M = (2 \text{ kips})(3 \text{ in}) = 6 \text{ kip} - \text{in}$$

$$\sigma = \frac{P}{A} + \frac{M}{S} = \frac{2 \text{ kips}}{0.785 \text{ in}^2} + \frac{6 \text{ kip} - \text{in}}{0.098 \text{ in}^3} = 63.77 \text{ ksi} \approx 64 \text{ ksi}$$

Reference: NCEES *PE Civil Reference Handbook* > Mechanics of Materials > Beams
Answer: C

49. Find vertical reaction at A and B:

$$R_{Ay} = \left(\frac{40 \text{ ft} - 25 \text{ ft}}{40 \text{ ft}}\right)(5{,}000 \text{ lb}) = 1{,}875 \text{ lb}$$

$$R_{By} = \left(\frac{25 \text{ ft}}{40 \text{ ft}}\right)(5{,}000 \text{ lb}) = 3{,}125 \text{ lb}$$

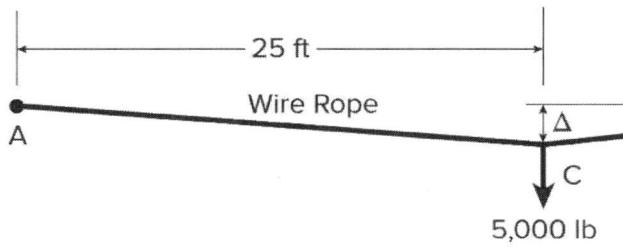

$$\Sigma M_C = R_{Ax}(3 \text{ ft}) - R_{Ay}(25 \text{ ft}) = 0$$

$$\therefore R_{Ax} = \frac{(1{,}875 \text{ lb})(25 \text{ ft})}{3 \text{ ft}} = 15{,}625 \text{ lb}$$

From the sum of horizontal forces:
$R_{Ax} = R_{Bx} = 15{,}625 \text{ lb}$
$\therefore$ Max wire rope force $R = \sqrt{(15{,}625 \text{ lb})^2 + (3{,}125 \text{ lb})^2} = 15.93 \text{ kips} \approx 16)$

Reference: NCEES *PE Civil Reference Handbook* > Statics > Statically Determinate Truss
Answer: D

SOLUTIONS

50. Convert the load from pounds per square foot to pounds per foot along the beam length by multiplying the load in pounds per square foot by the distance between beams. The distributed loads on the member are:

W_D = 8 psf (6 ft) = 48 plf
W_{Lr} = 20 psf (6 ft) = 120 plf
W_{W+} = 25 psf (6 ft) = 150 plf
W_{W-} = 75 psf (6 ft) = 450 plf

Ultimate positive distributed load (W_{u+}):
$1.4W_D$ = 1.4(48 plf) = 67.2 plf
$1.2W_D + 0.5W_{Lr}$ = 1.2(48 plf) + 0.5(120 plf) = 117.6 plf
$1.2W_D + 1.6W_{Lr} + 0.5W_{W+}$ = 1.2(48 plf) + 1.6(120 plf) + 0.5 (150 plf) = 324.6 plf
$1.2W_D + 1.0W_{W+} + 0.5W_{Lr}$ = 1.2(48 plf) + 1.0(150 plf) + 0.5(120 plf) = 267.6 plf

Ultimate negative distributed load (W_{u-}):
$0.9w_D + 1.0W_{W+}$ = 0.9(48 plf) + 1.0(-450 plf) = -406.8 plf

Thus,
The ultimate positive moment the beam experiences is:
$M_{u+} = W_{u+}L^2 / 8$ = (324.6 plf)(30 ft)² / 8 = 36.5 kip-ft
The deck is bracing the top flange of the beam; therefore, it is completely braced for positive moment.
The ultimate negative moment the beam experiences is:
$M_{u-} = W_{u-}L^2 / 8$ = (-406.8 plf)(30 ft)² / 8 = -45.8 kip-ft
With no bracing at the bottom flange, the unbraced length would be the entire span—30 ft.
Therefore, the negative moment exerted on the member would control design.

Reference: ASCE/SEI Standard 7-10: *Minimum Design Loads for Buildings and Other Structures* > Section 2.3.2

Answer: A

SOLUTIONS

51. Per ASCE/SEI 7-10, section 4.5.1, there are three loading conditions to consider:
1. A 200-lb point load applied in any direction at any point on the handrail or top rail
2. A 50-lb/ft load applied in any direction along the handrail or the top rail, which does not need to be considered concurrently with the load mentioned in item 1
3. A 50-lb load on any area not to exceed 12 × 12 in; does not need to be considered concurrently with items 1 and 2

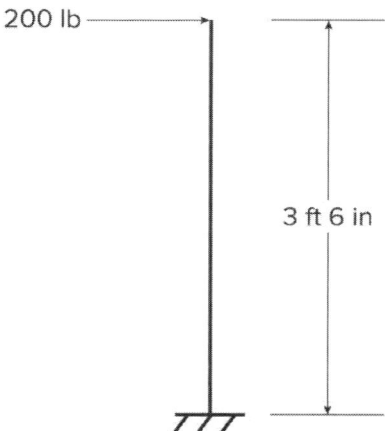

Since the loading conditions do not need to be considered concurrently, and with the 50-lb/ft load over the 3-ft tributary width resulting in a 150-lb resultant point load, the 200-lb load applied at the top rail in the post's unbraced direction, into the page, will create the controlling load effect.
Controlling load combination = $1.2D + 1.6$ (from ASCE/SEI 7-10, sect. 2.3.2)
With self-weight being neglected,
$M_u = 1.6Ph = 1.6(200 \text{ lb})(3.5 \text{ ft}) = 1{,}120$ lb-ft

Reference: ASCE/SEI Standard 7-10: *Minimum Design Loads for Buildings and Other Structures* > Section 4.5.1
Answer: D

SOLUTIONS

52. Determine center of rigidity in X and Y directions:

$$K = \frac{l^3}{3EI} \rightarrow \frac{1}{K} = R = \frac{3EI}{l^3}$$

Where:

$$E = 57,000\sqrt{f'_c} = \frac{57,000\sqrt{4,000}}{1,000} = 3,605 \text{ ksi}$$

$$I_{20\times20} = \frac{bh^3}{12} = 13,333 \text{ in}^4$$

$$I_{24\times24} = \frac{bh^3}{12} = 27,648 \text{ in}^4$$

$$R_{20} = \frac{3(3,605 \text{ ksi})(13,333 \text{ in}^4)}{(18 \text{ ft})^3} = 24,725 \text{ in}^2/\text{ft}^3$$

$$R_{24} = \frac{3(3,605 \text{ ksi}) \times 27,648 \text{ in}^4}{(18 \text{ ft})^3} = 51,271 \text{ in}^2/\text{ft}^3$$

$$\sum R = 4(R_{20}) + 5(R_{24}) = 355,255 \text{ in}^2/\text{ft}^3$$

Use gridline A as the reference axis to find $\bar{x}$:

$$\bar{x} = \frac{\Sigma R.x}{\Sigma R}$$

$$= \frac{[(2R_{20} + R_{24})(0 \text{ ft}) + 3(R_{24})(30 \text{ ft}) + 2(R_{20})(80 \text{ ft}) + (R_{24})(80 \text{ ft})]}{355,255 \text{ in}^2/\text{ft}^3}$$

$$= 35.67 \text{ ft}$$

Use gridline 3 as the reference axis to find $\bar{y}$:

$$\bar{y} = \frac{\Sigma R.y}{\Sigma R}$$

$$= \frac{[(2R_{20} + R_{24})(0 \text{ ft}) + 3(R_{24})(35 \text{ ft}) + 2(R_{20})(60 \text{ ft}) + (R_{24})(60 \text{ ft})]}{355,255 \text{ in}^2/\text{ft}^3}$$

$$= 32.2 \text{ ft}$$

Construct a table to determine polar moment of inertia:

COLUMN	R (in^2/ft^3)	d_x(ft)	d_y(ft)	$R(d_x^2 + d_y^2)$(in^2/ft)	$Rd_x/\Sigma Rr^2$(1/ft)
A1	24,725	35.67	27.8	50,567,296	
A2	51,271	35.67	2.8	65,636,565	
A3	24,725	35.67	32.2	57,094,696	
B1	51,271	5.67	27.8	41,272,586	
B2	51,271	5.67	2.8	2,050,271	
B3	51,271	5.67	32.2	54,808,130	
C1	24,725	44.33	27.8	67,696,776	
C2	51,271	44.33	2.8	101,157,114	0.00442
C3	24,725	44.33	32.2	74,224,176	
	$\Sigma R = 355,255$ in^2/ft^3			$\Sigma R(d_x^2 + d_y^2) =$ $\Sigma Rr^2 = 514,507,608$	

SOLUTIONS

Determine moment:
Accidental eccentricity = 80 ft (0.05) = 4 ft
Max moment will occur with the largest moment arm distance measured from the center of rigidity.
Distance between center of mass and center of rigidity = 40 ft $- \bar{x}$ = 4.33 ft
Total eccentricity, $e = 4.33$ ft $+ 4$ ft $= 8.33$ ft
$M_{Max} = Ve = (230 \text{ kips})(8.33 \text{ ft}) = 1{,}916 \text{ kip} - \text{ft}$
where:
$V = C_s W = 0.2(1{,}150 \text{ kips}) = 230 \text{ kips}$
Determine force in y-direction to column C-2.
$$F_y = \frac{VR_{C-2}}{\sum R} + \frac{MR_{C-2}d_x}{\sum Rr^2}$$
$$= (230 \text{ kips})\left(\frac{51{,}271 \text{ in}^2/\text{ft}^3}{355{,}255 \text{ in}^2/\text{ft}^3}\right) + 1{,}916 \text{ kip} - \text{ft}(0.00442/\text{ft})$$
$$= 33.2 + 8.5 = 41.7 \text{ kips}$$

Reference: ASCE/SEI Standard 7-10: *Minimum Design Loads for Buildings and Other Structures* > Section 12.14
Answer: D

SOLUTIONS

53. The obtained maximum considered ground acceleration should be modified based on site class. The modified accelerations are:

$S_{MS} = F_a S_s$ ASCE 7-10, Eq. 11.4-1

$S_{M1} = F_v S_1$ ASCE 7-10, Eq. 11.4-2

Where:

S_s = short-period acceleration or acceleration-based factor.

S_1 = long-period acceleration or velocity-based factor.

F_a = site coefficient from AISC 7-10, Table 11.4-1

F_v = site coefficient from AISC 7-10, Table 11.4-2

$F_a = 1.0 \qquad F_v = 1.3$

$S_{MS} = F_a S_s = 1(2.106) = 2.106 \text{ g}$

$S_{M1} = F_v S_1 = 1.3(0.761) = 0.989 \text{ g}$

The design spectral accelerations are:

$S_{DS} = \dfrac{2}{3} S_{MS}$ ASCE 7 − 10, Eq. 11.4 − 3

$S_{D1} = \dfrac{2}{3} S_{M1}$ ASCE 7 − 10, Eq. 11.4 − 4

$S_{DS} = \dfrac{2}{3}(2.106) = 1.404 \text{ g}$

$S_{D1} = \dfrac{2}{3}(0.989) = 0.659 \text{ g}$

The seismic coefficient from ASCE 7-10, Eq. 12.8-2 is:

$C_s = \dfrac{S_{DS}}{(R/I)}$

Where:

S_{DS} = design spectral acceleration for short period

R = response modification factor from ASCE 7-10, Table 12.2-1

I_e = seismic importance factor from AISC 7-10, Table 1.5-2

For steel special concentrically braced frames:

$R = 6$

For occupancy category II:

$I_e = 1.0$

$C_s = \dfrac{1.404}{(6/1)} = 0.234$

Lower limit: $C_s = 0.044 \, S_{DS} \, I_e \geq 0.01$ according to ASCE/SEI 7-10 Equation 12.8-5

$C_s = (0.044)(1.404 \text{ g})(1) = 0.062 \geq 0.01$

Upper limit: $C_s = \dfrac{S_{D1}}{T \dfrac{R}{I_e}}$ for $T \leq T_L$

SOLUTIONS

$$C_S = \frac{S_{D1} T_L}{T^2 \left(\frac{R}{I_e}\right)} \text{ for } T > T_L$$

$T = C_u T_a$ ASCE/SEI 7-10 section 12.8-2

$C_u = 1.4$ ASCE/SEI 7-10 Table 12.8-1

$T_a = C_t h_n^x$

$C_t = 0.02$ (all other structural systems) ASCE/SEI 7-10 Table 12.8-2

$x = 0.75$ (all other structural systems) ASCE/SEI 7-10 Table 12.8-2

$h_n = 15 + 12 + 12 = 39$ ft

$T_a = C_t h_n^x = 0.02(39^{0.75}) = 0.312$ s

$T = C_u T_a = 1.4(0.312) = 0.437$ s

$T_L = 8$ ASCE/SEI 7-10 Figure 22.12 for California

$$T \leq T_L \rightarrow C_{S-\text{Upper}} = \frac{S_{D1}}{T\left(\frac{R}{I_e}\right)} = \frac{0.659}{0.437 \left(\frac{6}{1}\right)} = 0.251$$

$C_{S-\text{Lower}} = 0.062 < C_S = 0.234 < C_{S-\text{Upper}} = 0.251 \rightarrow C_S = 0.234$

Total weight (W):

$W = 850 + 850 + 700 = 2{,}400$ kips

Seismic base shear:

$V = C_s W$ ASCE 7-10, Eq. 12.8-1

$V = 0.234(2{,}400 \text{ kips}) = 561.6$ kips (~ 562)

Reference: ASCE/SEI Standard 7-10: *Minimum Design Loads for Buildings and Other Structures* > Section 12.8.3

Answer: D

SOLUTIONS

54. The modulus of rupture f_r is the flexural tensile stress that causes cracking in a beam. Per ACI Code Eq. 19.2.3.1, the modulus of rupture is:

$$f_r = 7.5\left(\sqrt{f_c'}\right) = 7.5\left(\sqrt{4{,}000 \text{ psi}}\right) = 474 \text{ psi}$$

Equate the tensile stress at the beam extreme fiber $\left(\frac{Mc}{I}\right)$ to f_r and solve for the beam depth.

$$f_r = \frac{Mc}{I} = \frac{M\frac{h}{2}}{\frac{bh^3}{12}} = \frac{6M}{bh^2} = 474 \text{ psi}$$

Where:

$$c = \frac{h}{2}$$

$$\frac{(6)(80 \text{ k} - \text{ft})(12 \text{ in/ft})(1{,}000 \text{ lb/kip})}{(15)(h^2)} = 474 \text{ psi}$$

$$h = 28.5 \text{ in}$$

References: ACI Code 318-14: *Building Code Requirements for Structural Concrete and Commentary* > Section 19.2.3
Answer: D

55. According to the *Steel Manual*, corrosion protection is not required for steel that is enclosed by building finish, in contact with concrete, or coated with contact-type fireproofing.

Reference: AISC *Steel Manual* 14th Edition > General Design Considerations > Corrosion Protection
Answer: A, B, E

56. Splitting tensile strength = f_{ct}

$$f_{ct} = \frac{2P}{\pi DL} = \frac{2(51{,}000 \text{ lb})}{(\pi)(6 \text{ in})(12 \text{ in})} = 451 \text{ psi}$$

Compressive strength:

$$f_c' = \frac{P}{A} = \frac{127{,}560 \text{ lb}}{\pi \times 6^2/4} = 4{,}512 \text{ psi}$$

$$\text{Ratio} = \frac{451 \text{ psi}}{4{,}512 \text{ psi}} = 10\%$$

Reference: ACI Code 318-14: *Building Code Requirements for Structural Concrete and Commentary* > Section 2.3
Answer: A

SOLUTIONS

57. Create a diagram:

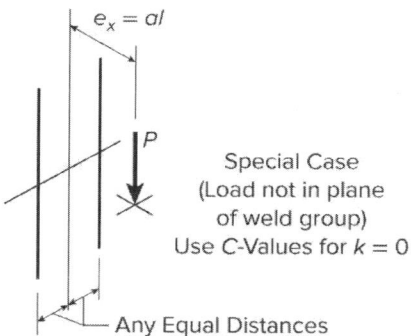

Using *Steel Manual* Table 8-4 with angle = 0°:
$P_u = 100$ kips
$k = 0$ (load not in plane of weld group)
$e_x = 18$ in
$l = 24$ in
$a = \dfrac{e_x}{l} = \dfrac{18 \text{ in}}{24 \text{ in}} = 0.75$

C value can be obtained according to Table 8-4.
$C = (1.76 + 1.57)/2 = 1.665$ kip/in for $k = 0$ and $a = 0.75$
$C_1 = 1.0$ electrode strength coefficient from *Steel Manual* Table 8-3 (1 for E70XX electrodes)
$\phi = 0.75$
D = number of sixteenths-of-an-inch in the fillet weld size
$D_{\text{Min}} = \dfrac{P_u}{\phi C C_1 l} = \dfrac{100 \text{ kips}}{0.75(1.665 \text{ kip/in})(1)(24 \text{ in})} = 3.3\ (\sim 4)$
$D = 4$

The calculated value is ¼ in, but according to *Steel Manual* Table J2-4, for the bracket with 1-in thickness, the minimum size of fillet weld is 5/16 in (controls).

References: AISC *Steel Manual* 14th Edition > Tables 8-3, 8-4 and J2-4
Answer: C

SOLUTIONS

58. $L_{Brace} = \sqrt{(6 \text{ ft})^2 + (8 \text{ ft})^2} = 10 \text{ ft}$

For condition 1:

At joint C:

$\Sigma F_x = 0 \qquad R_{x,c} + F_{Brace,x} = 0$

$F_{Brace} \left(\dfrac{8 \text{ ft}}{10}\right) = -10 \text{ kips} \rightarrow F_{Brace} = -12.5 \text{ kips}$

For condition 1, the brace is in compression.

Using *Steel Manual* Table 4-12:

L3 1/2 × 3 1/2 × 5/16 $\qquad \phi P_n = 11.5 \text{ kips}$
L3 1/2 × 3 1/2 × 3/8 $\qquad \phi P_n = 13.1 \text{ kips}$

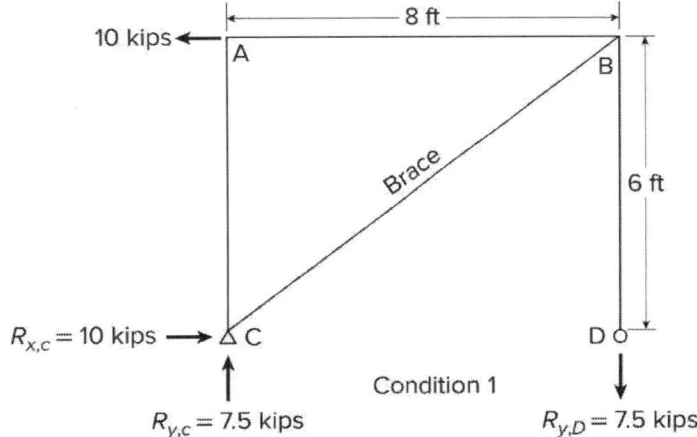

Condition 1

SOLUTIONS

For condition 2:
At joint C:
$\sum F_x = 0 \qquad -R_{x,c} + F_{Brace,x} = 0$

$F_{Brace}\left(\dfrac{8\text{ ft}}{10}\right) = 30\text{ kips} \;\rightarrow\; F_{Brace} = 37.5\text{ kips}$

For condition 2, brace is in tension.

Using *Steel Manual* Table 5-2:
Yielding failure controls:
L3 1/2 × 3 1/2 × 1/4 $\qquad \phi P_n = 55.1$ kips
L3 1/2 × 3 1/2 × 5/16 $\qquad \phi P_n = 68.0$ kips

The compression condition controls.

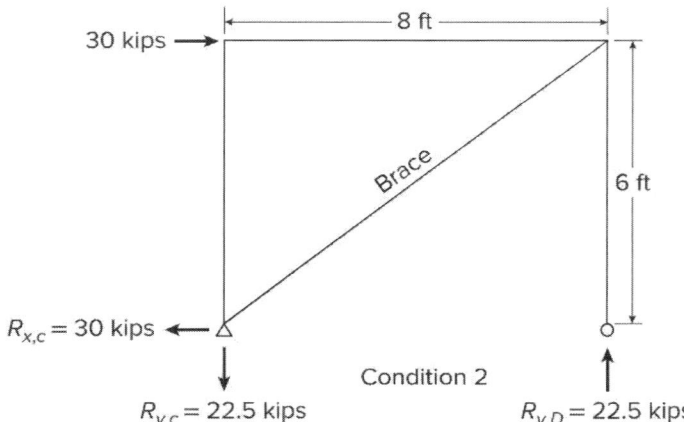

Condition 2

Reference: AISC *Steel Manual* 14th Edition> Design of Compression Members > Eccentrically Loaded Single Angles in Axial Compression
Answer: C

SOLUTIONS

59. $A_g = (6 \text{ in})(6 \text{ in}) = 36 \text{ in}^2$
$A_{st} = (0.6 \text{ in}^2)(4 \text{ bars}) = 2.4 \text{ in}^2$
$f'_c = 4{,}000 \text{ psi}$
$f_y = 60{,}000 \text{ psi}$
Tied column; use $0.80 P_o$ (ACI Table 22.4.2.1)
$P_n = 0.80[0.85 f'_c (A_g - A_{st}) + f_y A_{st}]$
$= 0.80[0.85(4{,}000 \text{ psi})(36 \text{ in}^2 - 2.4 \text{ in}^2) + (60{,}000 \text{ psi})(2.4 \text{ in}^2)]$
$= 206{,}592 \text{ lb} = 206.6 \text{ kips}$
$\phi = 0.65$ (tied column)
$\phi P_n = (0.65)(206.6 \text{ kips}) = 134.3 \text{ kips}$

Reference: ACI Code 318-14: *Building Code Requirements for Structural Concrete and Commentary* > Chapter 22 > Section 22.4
Answer: B

60. $A = (B)(L) = (10 \text{ ft})^2 = 100 \text{ ft}^2$
$W_{\text{Footing}} = (A)(t)(\gamma_c) = (100 \text{ ft}^2)(1.5 \text{ ft})(0.150 \text{ kcf}) = 22.5 \text{ kips}$
$P_u = 1.2(50 \text{ kips} + 22.5 \text{ kips}) + 1.6(85 \text{ kips}) = 223 \text{ kips}$
$q_u = P_u/A = 223 \text{ kips}/100 \text{ ft}^2 = 2.23 \text{ ksf}$
The critical section for flexure is at the face of the column.
$f = L/2 - c/2 = 10 \text{ ft}/2 - 3 \text{ ft}/2 = 3.5 \text{ ft}$
$M_u = q_u L(f)(f/2) = (2.23 \text{ ksf})(10 \text{ ft})(3.5 \text{ ft})(3.5 \text{ ft}/2) = 136.6 \text{ kip} - \text{ft}$

Reference: NCEES *PE Civil Reference Handbook* > Structural Analysis > Moment, Shear, and Deflection Diagrams
Answer: C

SOLUTIONS

61. n = number of piles = 12
The horizontal pile spacing:
$$s_x = \frac{12 \text{ ft}}{3} = 4 \text{ ft}$$

The vertical pile spacing:
$$s_y = \frac{9 \text{ ft}}{2} = 4.5 \text{ ft}$$

Moment of inertia of pile group about x-axis:
$$I_x = 2 \times 4 \times (s_y)^2 = 2(4)(4.5 \text{ ft})^2 = 162 \text{ ft}^2$$

Moment of inertia of pile group about y-axis:
$$I_y = 2(3)\left[\left(\frac{s_x}{2}\right)^2 + \left(\frac{s_x}{2} + s_x\right)^2\right]$$
$$= 2(3)\left[\left(\frac{4 \text{ ft}}{2}\right)^2 + \left(\frac{4 \text{ ft}}{2} + 4 \text{ ft}\right)^2\right] = 240 \text{ ft}^2$$

Horizontal distance from pile group centroid to extreme pile:
$$c_x = 6 \text{ ft}$$

Vertical distance from pile group centroid to extreme pile:
$$c_y = 4.5 \text{ ft}$$

The maximum force in extreme pile:
$P_{\text{Max}} = P/n + M_x c_y/I_x + M_y c_x/I_y$
$= 300 \text{ kips}/12 + (500 \text{ kip} - \text{ft})(4.5 \text{ ft})/162 \text{ ft}^2 + (350 \text{ kip} - \text{ft})(6 \text{ ft})/$
$240 \text{ ft}^2 = 47.65 \text{ kips}$

Reference: NCEES *PE Civil Reference Handbook* > Statics > Systems of Forces
Answer: C

SOLUTIONS

62. Solve for the critical section.

$$d = t - c_c - \frac{d_{Bar}}{2} = 2\text{ ft} - \left(\frac{3\text{ in}}{12\text{ in/ft}}\right) - \frac{\frac{1.27\text{ in}}{12\text{ in/ft}}}{2} = 1.70\text{ ft}$$

$$b_o = 4(c + d) = 4(2\text{ ft} + 1.7\text{ ft}) = 14.8\text{ ft}$$

Per ACI Code 318-14 Table 22.6.5.2, calculate the controlling value of V_c.
$\lambda = 1$ (normal weight concrete)
$f'_c = 3{,}500$ psi
$V_c = 4\lambda\sqrt{f'_c}b_o d$
$= (4)(1)\sqrt{3{,}500\text{ psi}}\,(14.8\text{ ft})(12\text{ in/ft})(1.7\text{ ft})(12\text{ in/ft})\dfrac{\text{kips}}{1{,}000\text{ lb}}$
$= 857$ kips

$\beta = 1$ (square column)
$V_c = (2 + 4/\beta)\lambda\sqrt{f'_c}b_o d$
$= (2 + 4/1)(1)\sqrt{3{,}500\text{ psi}}(14.8\text{ ft})(12\text{ in/ft})(1.7\text{ ft})(12\text{ in/ft})\dfrac{\text{kips}}{1{,}000\text{ lb}}$
$= 1{,}286$ kips

$\alpha_s = 40$ (interior column)
$V_c = (2 + \alpha_s d/b_o)\sqrt{f'_c}b_o d$
$= (2 + 40(1.7\text{ ft})/14.8\text{ ft})\sqrt{3{,}500\text{ psi}}(14.8\text{ ft})(12\text{ in/ft})(1.7\text{ ft})(12\text{ in/ft})\dfrac{\text{kips}}{1{,}000\text{ lb}}$
$= 1{,}413$ kips

Select the smallest value, so V_c is 857 kips
$\phi = 0.75$
$\phi V_n = 0.75(857\text{ kips}) = 642.75$ kips

Reference: ACI Code 318-14: *Building Code Requirements for Structural Concrete and Commentary* > Section 22.6
Answer: B

SOLUTIONS

63. F_c = 800 psi (Table 4D)
C_D = 1 (Table 2.3.2)
C_M = 1 (4.3.3)
C_t = 1 (Table 2.3.3)
C_F = 1 (4.3.6)
C_i = 1 (4.3.8)

Calculate C_p:
C_T = 1 (4.4.2)
E_{Min} = 400,000 psi (Table 4D)
$E_{Min}' = E_{Min} C_M C_t C_i C_T$ = (400,000 psi)(1)(1)(1)(1) = 400,000 psi
l_e = 300 in
d = 13.5 in (Table 1B)
$F_{cE} = 0.822 E_{Min}'/(l_e/d)^2$ = 0.822(400,000 psi)/(300/13.5)^2 = 666 psi
$F_c^* = F_c C_D C_M C_t C_F C_i$ = (800 psi)(1)(1)(1)(1)(1) = 800 psi
F_{cE}/F_c^* = 666 psi/800 psi = 0.833
c = 0.8
Use Equation (3.7.1) for C_p

$$C_p = \frac{1 + \left(\frac{F_{cE}}{F_c^*}\right)}{2c} - \sqrt{\left[\frac{1 + \frac{F_{cE}}{F_c^*}}{2c}\right]^2 - \frac{\frac{F_{cE}}{F_c^*}}{c}}$$

C_p = 0.63
l_e/d = 22 < 50 (3.7.1.4)
$F'_c = F_c C_D C_M C_t C_F C_i C_p$ = (800 psi)(1)(1)(1)(1)(1)(0.63) = 504 psi
A = 182.3 in^2 (Table 1B)
Axial capacity = (504 psi)(182.3 in^2) = 91.9 kips

References: AWC *National Design Specification for Wood Construction*, 2015 > Section 3.6
Answer: B

SOLUTIONS

64. First, calculate eccentricity:

$$e = \frac{M}{P} = \frac{81 \text{ kip} - \text{ft} \times 12 \text{ in/ft}}{500 \text{ kips}} = 1.94 \text{ in}$$

$$e_{\text{Kern}} = \frac{L}{6} = \frac{14 \text{ ft} \times 12 \text{ in/ft}}{6} = 28 \text{ in}$$

$e < e_{\text{Kern}}$; therefore:

$$q_{\text{Max}} = \frac{P}{BL} + \frac{6Pe}{BL^2}$$

Where:
B = small dimension; 8 ft
L = large dimension; 14 ft

$$q_{\text{Max}} = \frac{500 \text{ kips}}{(8 \text{ ft})(14 \text{ ft})}\left[1 + \frac{6(1.94 \text{ in})\left(\frac{1 \text{ ft}}{12 \text{ in}}\right)}{14 \text{ ft}}\right] = 4.8 \text{ ksf}$$

Reference: NCEES *PE Civil Reference Handbook* > Bearing Capacity > Bearing Capacity Equation for Concentrically Loaded Strip Footings
Answer: C

65. The tension face in this wall is the outer face.

Reference: ACI Code 318-14: *Building Code Requirements for Structural Concrete and Commentary* > Chapter 11
Answer: point A

SOLUTIONS

66. The capacity of a slip-critical bolt group is determined in accordance with AISC Specification sections J3.8 and J3.9.

Bolt tensile capacity:
Per AISC Specification Table J3.2 for group A bolts, A325, the nominal tensile strength is F_{nt} = 90 ksi.

Thus, the tensile capacity is:
$\phi R_n = \phi F_{nt} A_b = 0.75(90 \text{ ksi})(0.785 \text{ in}^2) = 53 \text{ kips/bolt}$

Bolt slip resistance:
The slip resistance capacity of a bolt is determined using AISC Specification section J3.8.
$\phi R_n = \phi \mu D_u h_f T_b n_s$ (Eqn. J3 − 4)
$\mu = 0.30$ for class A surfaces
$D_u = 1.13$ (multiplier that reflects the ratio of mean installed bolt pretension to the minimum bolt pretension)
$h_f = 1.0$ (filler factor; assume no more than one filler)
$T_b = 51$ kips (per AISC Table J3.1)
$n_s = 1$ (number of slip planes)
$\phi = 1.00$ (for standard size holes)
$\phi R_n = 1.00(0.30)(1.13)(1.0)(51)(1.00) = 17.3 \text{ kips/bolt}$

Total slip resistance of connection:
Because the connection is subjected to tension and shear, a combined effect must be considered. The capacity can be determined via AISC section J3.9.
$n_b = 6$ (corresponds to the number of bolts carrying the applied tension)
The combined slip-critical tension and shear coefficient is:
$k_{sc} = 1 - \dfrac{T_u}{D_u T_b n_b} = 1 - \dfrac{(50 \text{ kips})}{(1.13)(51 \text{ kips})(6)} = 0.855$

Therefore, the total capacity would be:
$\phi R_n = 1.00(17.3 \text{ kips/bolt})(6 \text{ bolts})(0.855) = 88.75 \text{ kips}$

Reference: AISC *Steel Manual* 14th Edition > Specifications and Codes > Section 16.1 J3
Answer: B

SOLUTIONS

67. Get material properties from NDS 2015 Supplement Table 4A for Spruce-Pine-Fir #2:

$F_c = 1{,}150$ psi
$E_{Min} = 510{,}000$ psi
$b = 1.5$ in and $d = 5.5$ in (NDS Table 1B)
$A = 8.25$ in² (NDS Table 1B)

Calculate C_p per NDS Specification 2015 section 3.7.1
$C_F = 1.1$ (NDS 2015 Supplement for F_c and $d = 6$ in)
$F_c^* = 1{,}150$ psi $(1.1) = 1{,}265$ psi

$$F_{CE} = \frac{0.822 E'_{Min}}{\left(\frac{l}{d}\right)^2} = 880 \text{ psi}$$

Where:
$E'_{Min} = E_{Min}$ (all adjustment factors are 1)
$K_e l = l_e = 1.0(10 \text{ ft} \times 12 \text{ in/ft}) = 120$ in
$\frac{l_e}{d} = \frac{120 \text{ in}}{5.5 \text{ in}} = 21.8$
$\frac{l_e}{d} < 50$
$c = 0.8$ for sawn lumber

$$C_p = \frac{1 + \left(\frac{F_{CE}}{F_c^*}\right)}{2c} - \sqrt{\left[\frac{1 + \left(\frac{F_{CE}}{F_c^*}\right)}{2c}\right]^2 - \frac{\frac{F_{CE}}{F_c^*}}{c}}$$

$$\left(\frac{F_{CE}}{F_c^*}\right) = \left(\frac{880 \text{ psi}}{1{,}265 \text{ psi}}\right) = 0.6956$$

$$C_p = \frac{1 + (0.6956)}{2(0.8)} - \sqrt{\left(\left[\frac{1 + (0.6956)}{2(0.8)}\right]\frac{1 + (0.6956)}{2(0.8)}\right)^2 - \frac{0.6956}{0.8}} = 0.556$$

Calculate allowable vertical load:
$F_c' = F_c^* C_p = (1{,}265 \text{ psi})(0.556) = 703$ psi
Allowable stud vertical load = $P = F_c' A = (703 \text{ psi})(8.25 \text{ in}^2) = 5{,}800$ lb/stud
Determine required number of studs:
$$\frac{10{,}000 \text{ lb}}{(5{,}800 \text{ lb/stud})} = 1.73 \text{ studs } (\sim 2)$$

Reference: NDS Supplement *Design Values for Wood Construction* 2015 Edition > Table 4A
Answer: B

SOLUTIONS

68. Determine limits per TMS 402-13 or ACI 530-13:
$M = 50 \text{ kips}(12 \text{ ft}) = 600 \text{ kip} - \text{ft}$
$V = 50 \text{ kips}$
$d_v = 8 \text{ ft}$
$\gamma_g = 1.0$
$\dfrac{M}{V d_v} = \dfrac{600 \text{ kip} - \text{ft}}{(50 \text{ kips})(8 \text{ ft})} = 1.5 > 1.0$
$\rightarrow F_{v,\text{Max}} = \left(2\sqrt{f'_m}\right)\gamma_g = (2\sqrt{1{,}500 \text{ psi}})(1) = 77.4 \text{ psi}$

Solve for applied shear stress:
$f_v = \dfrac{V}{A_{nv}} = \dfrac{50{,}000 \text{ lb}}{(7.625 \text{ in})(8 \text{ ft})(12 \text{ in/ft})} = 68.3 \text{ psi}$
$f_v < F_{v,\text{Max}}$

Solve for allowable masonry shear stress. (Note: $\dfrac{M}{Vd_v}$ shall not exceed 1.0 for the calculation of allowable masonry shear stress.)

$F_{vm} = \dfrac{1}{2}\left[\left(4.0 - 1.75\left(\dfrac{M}{Vd_v}\right)\right)\sqrt{f'_m}\right] \rightarrow$ ACI 530 (Eqn. 8 − 29); use $\dfrac{M}{Vd_v} = 1.0$

$\therefore F_{vm} = \dfrac{1}{2}\left[(4.0 - 1.75(1))\sqrt{1{,}500 \text{ psi}}\right] = 43.5 \text{ psi}$
$f_v > F_{vm}$
$\therefore$ Steel reinforcement is required.

Solve for shear stress resisted by shear reinforcing steel and then solve for required reinforcement spacing:
$F_{vs} = f_v - F_{vm} = 68.3 \text{ psi} - 43.5 \text{ psi} = 24.8 \text{ psi}$
Rearranging ACI 530 Equation 8 − 30 yields: $s = 0.5 \left(\dfrac{A_v F_s d_v}{A_{nv} F_{vs}}\right)$
$A_v = 0.2$ for one #4 bar
$s = 0.5 \left(\dfrac{[(0.2 \text{ in}^2)(32{,}000 \text{ ksi})(8 \text{ ft})(12 \text{ in/ft})]}{(7.625 \text{ in})(8 \text{ ft})(12 \text{ in/ft})(24.8 \text{ psi})}\right) = 16.9 \text{ in}$

Reference: TMS 402/601-2013 > Allowable Stress Design of Masonry > Section 8.3.5.1.2
Answer: B

SOLUTIONS

69. Determine composite section properties:
Area of one #4 bar: $A_b = 0.2$ in²
$A_s = A_b(2) = 0.2 \text{ in}^2 (2) = 0.4 \text{ in}^2$
$d = 14$ in
$b = 7.625$ in

$$\rho = \frac{A_s}{bd} = \frac{(0.4 \text{ in}^2)}{(7.625 \text{ in})(14 \text{ in})} = 0.00375$$

$E_m = 900 F'_m = 900 \times 1{,}500 \text{ psi} = 1{,}350{,}000 \text{ psi or } 1{,}350 \text{ ksi}$

$$n = \frac{E_s}{E_m} = \frac{29{,}000 \text{ ksi}}{1{,}350 \text{ ksi}} = 21.5$$

$n\rho = 21.5(0.00375) = 0.080625$
$k = \sqrt{(n\rho)^2 + 2(n\rho)} - n\rho$
$k = \sqrt{(0.080625)^2 + 2(0.080625)} - 0.080625 = 0.329$
$j = 1 - \frac{k}{3} = 1 - \frac{0.329}{3} = 0.89$

Check masonry strength and steel strength in bending.
Allowable masonry stress: $F_b = 0.45 f'_m = 0.45 (1{,}500 \text{ psi}) = 675$ psi

$$F_b = \frac{2M}{b.d^2 j.k}$$

Then the masonry moment:

$$M_m = \frac{(F_b.b.d^2.j.k)}{2} = \frac{(675 \text{ psi})(7.625)(14 \text{ in})^2 (0.89)(0.329)}{2}$$
$= 147{,}692 \text{ lb} - \text{in or } 12.3 \text{ k} - \text{ft}$

$$F_s = \frac{M}{A_s.j.d}$$

Then the steel moment:
$M_s = F_s.A_s.j.d = (32 \text{ ksi})(0.4 \text{ in}^2)(0.89)(14 \text{ in})$
$\qquad\qquad = 159 \text{ kip} - \text{in or } 13.3 \text{ kip} - \text{ft}$
$M_m = 12.3 \text{ kip} - \text{ft} \leftarrow$ Controls

Reference: ACI 530-13 > Strength Design of Masonry > Chapter 9
Answer: C

SOLUTIONS

70. K = effective length factor (K = 1 for pinned-pinned end condition)
L = unbraced length of the column

Calculate the controlling slenderness ratio:
$$\left(\frac{KL}{r}\right)_x = \frac{(1)(10 \text{ ft})(12 \text{ in/ft})}{2.3 \text{ in}} = 52.2$$
$$\left(\frac{KL}{r}\right)_y = \frac{(1)(10 \text{ ft})(12 \text{ in/ft})}{1.78 \text{ in}} = 67.4$$

The buckling will be about the y-axis (weak axis) since the slenderness ratio about the y-axis is larger than the slenderness ratio about the x-axis.
Check the critical slenderness ratio with the slenderness ratio limit:
$$\left(\frac{KL}{r}\right)_y \leq 4.71\sqrt{\frac{E}{F_y}} = 4.71\sqrt{\frac{29{,}000 \text{ ksi}}{50 \text{ ksi}}} = 113.4$$

Since 67.4 < 113.4, therefore, inelastic buckling, use *Steel Manual* Equation E3-2:
$$F_{cr} = \left[(0.658^{\frac{F_y}{F_e}})\right]F_y$$
F_e = elastic buckling stress from *Steel Manual* Equation E3-4
$$F_e = \frac{\pi^2 E}{\left(\frac{KL}{r}\right)_y^2}$$
$$F_e = \frac{\pi^2 29{,}000}{(67.4)^2} = 63 \text{ ksi}$$
$$F_{cr} = \left[0.658^{\frac{F_y}{F_e}}\right]F_y = \left[0.658^{\frac{50}{63}}\right](50) = 35.87 \text{ ksi}$$
$$P_{cr} = F_{cr} A_g = (35.87 \text{ ksi})(18.9 \text{ in}^2) = 677.9 \text{ kips}$$

Reference: AISC *Steel Manual* 14th Edition > Specifications and Codes > Section 16.1 E3
Answer: A

SOLUTIONS

71. The distance between the load and center of the bolt group = eccentricity, e:

$e = 7.5$ in $+ 3$ in$/2 = 9$ in

$M = Pe = (20$ kips$)(9$ in$) = 180$ k-in

The table below provide the x and y coordinates for each bolt (d_{xi} and d_{yi}, respectively) and the distance between each bolt center and the bolt group centroid, d_i.

$M = 9$ in$(20$ kips$) = 180$ kips–in

BOLT	d_{xi}	d_{yi}	d_i	$(d_i)^2$
1	1.5	4.5	4.74	22.47
2	1.5	4.5	4.74	22.47
3	1.5	1.5	2.12	4.5
4	1.5	1.5	2.12	4.5
5	1.5	1.5	2.12	4.5
6	1.5	1.5	2.12	4.5
7	1.5	4.5	4.74	22.47
8	1.5	4.5	4.74	22.47
			$\Sigma d_i^2 =$	107.88 in^2

$J = \Sigma d_i^2 = 107.88$ in²

Shear force due to applied load P:

$$R_{yP} = \frac{P}{\text{\# of bolts}} = \frac{20 \text{ kips}}{8} = 2.5 \text{ kips} \downarrow$$

Bolts 1, 2, 7, and 8 are the farthest from the bolt group center of gravity since the d_i for each is the maximum.

Shear force components due to applied moment, $M = Pe = 180$ kip-in:
Vertical component:

$$R_{yM} = \frac{M \times d_{xi}}{J} = \frac{(180 \text{ kip} - \text{in})(1.5 \text{ in})}{107.88 \text{ in}^2} = 2.5 \text{ kips}$$

Horizontal component:

$$R_{xM} = \frac{M \times d_{yi}}{J} = \frac{(180 \text{ kip} - \text{in})(4.5 \text{ in})}{107.88 \text{ in}^2} = 7.5 \text{ kips}$$

Maximum resultant shear forces are in bolts 2 and 8 because the vertical components due to P and M are in the in same direction (downward) for these bolts.

Total vertical component in bolts 2 and 8:
$R_y = R_{yP} + R_{yM} = 2.5$ kips $\downarrow + 2.5$ kips $\downarrow = 5$ kips $\downarrow$

SOLUTIONS

Horizontal component in bolts 2 and 8:
$R_x = R_{xM} = 7.5$ kips

Resultant force:
$$R_{\text{Max}} = \sqrt{(R_x)^2 + (R_y)^2} = \sqrt{(7.5 \text{ kips})^2 + (5 \text{ kips})^2} = 9 \text{ kips}$$

Reference: AISC *Steel Manual* 14th Edition > Design Considerations for Bolts > Table 7-6
Answer: D

72. Shear lag factor U from AISC Table D3.1 is:
$U = 0.87 \qquad 2w > L \geq 1.5w$
Where:
Width of flat bar (w) = 6.5 in
Length of weld (L) = 12 in
$2(6.5 \text{ in}) = 13 \text{ in} > 12 \text{ in} \geq 1.5(6.5 \text{ in}) = 9.75 \text{ in}$

Shear lag factor from AISC Table D3.1 Case 4, $U = 0.87$
$A_g = wt$
$A_g = (6.5 \text{ in})\left(\dfrac{5}{8} \text{ in}\right) = 4.0625 \text{ in}^2$
$A_n = A_g$
$A_e = U A_n$
$A_e = 0.87(4.0625 \text{ in}^2) = 3.53 \text{ in}^2$

The effective net area is 3.53 in^2.

References: AISC *Steel Manual* 14th Edition > Specifications and Codes > Section 16.1 Table D3.1
Answer: B

SOLUTIONS

73. First, the relative stiffness of A and B must be determined. Because B is a fixed support, $G_B = 1$.

$$G_A = \frac{\sum \frac{I_{Col}}{L_{Col}}}{\sum \frac{I_{Beam}}{L_{Beam}}}$$

Both the columns above and below joint A are W12 × 58
From AISC Table 1-1, I_x for these columns = 475 in⁴
The column lengths are different as shown in the figure (10 ft for the first story and 12 ft for the second story).
Both girders adjoining joint A are W14 × 132 with I_x = 1,530 in⁴ (from AISC Table 1-1) and with different lengths of 18 ft and 22 ft.

$$G_A = \frac{\frac{475 \text{ in}^4}{10 \text{ ft}} + \frac{475 \text{ in}^4}{12 \text{ ft}}}{\frac{1{,}530 \text{ in}^4}{18 \text{ ft}} + \frac{1{,}530 \text{ in}^4}{22 \text{ ft}}} = 0.56$$

From the alignment chart AISC Fig. C-A-7.1, for sideway inhibited case (because of bracing), $K = 0.65$.

References: AISC *Steel Manual* 14th Edition > Dimensions and Properties > Table 1-1
AISC *Steel Manual* 14th Edition > Specifications and Codes > Section 16.1 > Appendix 7 Table C-A-7.1
Answer: A

74. For W14 × 132: $A = 38.8 \text{ in}^2, r_x = 6.28 \text{ in}, r_y = 3.76 \text{ in}$ (from AISC Table 1-1)
A36 is used, so $F_y = 36$ ksi.
Recommended design values for K: $K_x = 1, K_y = 0.8$ (from AISC Table C-A-7.1)

$$\frac{K_x L}{r_x} = \frac{1(60 \text{ ft})(12 \text{ in/ft})}{6.28 \text{ in}} = 114.6$$

$$\frac{K_y L}{r_y} = \frac{0.8(60 \text{ ft})(12 \text{ in/ft})}{3.76 \text{ in}} = 153.2 \text{ (governs)}$$

From Steel Manual Table 4 - 22 for $F_y = 36$ ksi and $KL/r = 153$, $\theta F_{cr} = 9.65$ ksi
Force: $\theta P_{cr} = A\theta F_{cr} = (38.8 \text{ in}^2)(9.65 \text{ ksi}) = 374.4$ kips

Reference: AISC *Steel Manual* 14th Edition > Dimensions and Properties > Table 1-1
AISC *Steel Manual* 14th Edition > Specifications and Codes > Section 16.1 > Appendix 7 Table C-A-7.1
Answer: B

SOLUTIONS

75. Use ACI Code 318-14 Equation 6.6.4.5.2 to calculate the magnification factor for non-sway columns.

$$\delta = \frac{C_m}{1 - \frac{P_u}{0.75 P_c}}$$

Columns with transverse loads applied between support C_m = 1.0 (ACI Code 318-14 sect. 6.6.4.5.3b):
P_u = Factored load = Max {1.4D, 1.2D + 1.6L} (ACI Code 318-14 Table 5.3.1)
P_u = Max {1.4 × 30, 1.2 × 30 + 1.6 × 25} = Max {42, 76} = 76 kips
P_c = 210 kips

$$\delta = \frac{1}{1 - \frac{76 \text{ kips}}{0.75(210 \text{ kips})}} = 1.93 \geq 1$$

Reference: ACI Code 318-14: *Building Code Requirements for Structural Concrete and Commentary* > Section 6.6.4
Answer: B

76. b_{Eff} = effective char rate (in/hr) adjusted for time, t
β_n = nominal char rate (in/hr), linear char rate based on 1-hour exposure
t = exposure time (hr)

$$\beta_{Eff} = \frac{1.2\, \beta_n}{t^{0.187}} \quad (NDS - 2015 \text{ Eqn. } 16.2 - 1)$$

$$\beta_{Eff} = \frac{1.2 \times 1.5}{2.25^{0.187}} = 1.55 \text{ in/hr}$$

Reference: AWC *National Design Specification for Wood Construction*, 2015 > Chapter 16 > Table 16.2.1A
Answer: B

77. In this case, the cantilevered section of roadway would experience vehicular and pedestrian loads. Thus, according to AASHTO, the deflection limit for the cantilevered section would be Span/375.

Reference: AASHTO *LRFD Bridge Design Specifications*, 7th Edition > Section 2.5.2.6.2
Answer: D

SOLUTIONS

78. According to the vectors' law:

$T = \sqrt{F_x^2 + F_y^2 + F_z^2}$ = Tension in each corner cable

The equation of equilibrium is calculated as:
Weight of prefabricated building = 2 tons × 2,000 lbs/ton = 4,000 lb
Each corner cable has a vertical component F_z. There are four vertical components (four corners), so:
$\Sigma F_z = 0$; $4F_z = 4,000$ lb
$F_z = 1,000$ lb

The corner cable has the following components (projections):
$\frac{40 \text{ ft}}{2} = 20$ ft in x-direction, $\frac{8 \text{ ft}}{2} = 4$ ft in y-direction and 12 ft in z-direction.
So, the length of each corner cable $(L) = \sqrt{(20 \text{ ft})^2 + (4 \text{ ft})^2 + (12 \text{ ft})^2} = 23.66$ ft

Since $F_z = \frac{z}{L}T$, the cable force (T) can be calculated as:
$T = \frac{L}{z}F_z = \left(\frac{23.66 \text{ ft}}{12 \text{ ft}}\right)(1,000 \text{ lb}) = 1.971(1,000 \text{ lb}) = 1,971$ lb

Reference: NCEES *PE Civil Reference Handbook* > Statics > Systems of Forces
Answer: B

SOLUTIONS

79. Each cable should carry half the load, but the lining cable will increase the force inside the cable. So, according to vector's law:

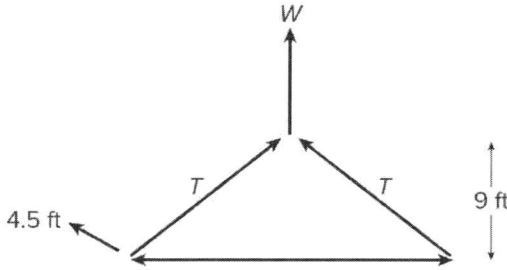

$2T \cos\theta = W = 4{,}000$ lb

The measure of the angle between T and the vertical axis is 26.5° because:
$\tan^{-1}(4.5 \text{ ft} / 9 \text{ ft}) = 26.565°$

$$T = \frac{4{,}000 \text{ lb}}{2 \cos 26.56} = 2{,}235.96 \text{ lb}$$

Applying a safety factor of 2:
$T = 4{,}471.9$ lb

Reference: NCEES *PE Civil Reference Handbook* > Statics > Systems of Forces
Answer: D

80. According to OSHA, a should be at least 3 ft.
Also:

$$b = \frac{(\text{Distance to top surface})}{4} = \frac{X}{4} = \frac{15 \text{ ft}}{4} = 3.75 \text{ ft}$$

Reference: OSHA Regulation 1910 > Subpart D
Answer: 3.75

You may have taken this practice test and studied the solutions, but what else can you do to thoroughly prepare for exam day?

Keep practicing!

One of the best study tools we offer is our Practice Portal Pro, which contains a bank of practice problems and solutions that closely mimics the NCEES' computer-based test (CBT) experience. There, you will be able to see and work through problems in the same format and style as your exam. The system will provide detailed answers so you can cross-reference your work and pinpoint where you may need improvement. Our goal with the Practice Portal Pro is to help you study smarter, not harder.

To gain immediate access to this comprehensive and time-saving study tool, download the free TotalAR app as explained at the front of this book. Use the app to scan the TAR code below and purchase the Practice Portal Pro (a $390 value).

This extremely useful tool will help you improve knowledge retention and gain a better understanding of the material.

www.schoolofpe.com